SpringerBriefs in Astronomy

For further volumes:
http://www.springer.com/series/10090

Michael Hoskin

William and Caroline Herschel

Pioneers in Late 18th-Century Astronomy

Springer

Royal Astronomical Society
Quicquid nitet notandum
1820
Advancing Astronomy and Geophysics

Michael Hoskin
Churchill College
University of Cambridge
Cambridge
UK

ISSN 2191-9100 ISSN 2191-9119 (electronic)
ISBN 978-94-007-6874-1 ISBN 978-94-007-6875-8 (eBook)
DOI 10.1007/978-94-007-6875-8
Springer Dordrecht Heidelberg New York London

Library of Congress Control Number: 2013939048

Printed on acid-free paper

Springer is part of Springer Science+Business Media (www.springer.com)

Foreword

When John Keats wrote, “Then felt I like some watcher of the skies When a new planet swims into his ken”, his literary allusion was to William Herschel, musician, composer, and conductor, who in 1781 discovered the planet that would be named Uranus. Herschel, who as an amateur astronomer had built the finest telescope then available, did what no other person alive could have done: he recognized something unusual when a peculiar bluish object with a distinct disk swam into his ken. No other person would have noticed it because his was the only telescope precise enough to show the disk-like structure. Herschel himself was more sanguine about his discovery. “It has generally been supposed that it was a lucky accident that brought this star to my view”, he remarked in a manuscript autobiography. “This is an evident mistake. In the regular manner (that) I examined every star of the heavens not only of that magnitude but many far inferior it was that night *its turn* to be discovered.... Had business prevented me that evening, I must have found it the next, and the goodness of my telescope was such that I perceived its visible planetary disk as soon as I looked at it.”

Herschel supposed his object to be a peculiar comet. Only later, after other astronomers computed its orbit, was it deemed to be the first new planet to be known since antiquity. It was a life-changing opportunity for William. The transformation was from a fashionable musician whose symphonies are still played to becoming the King’s astronomer. It was the choice between competence and genius.

Herschels genius was manifested in many ways. As a telescope maker his talents were unsurpassed as he built ever larger reflectors. Eventually they were coveted by crowned heads throughout Europe. But he did not just build telescopes for others. The best he kept for his own well-targeted observational projects. He almost single-handedly built an entirely new astronomical discipline, the physical arrangement of the heavens, the way the stars were arranged in a gigantic grindstone-like structure. The professional astronomers of his day viewed the heavens chiefly as a two-dimensional map against which the planets and comets marched in their stately paths. Herschel added a third dimension, their distances, or even a fourth dimension as he began to ponder their structure in time as the heavens slowly evolved.

When William moved from the fashionable musical city of Bath to the rural countryside near the King's Windsor castle, he uprooted his little sister Caroline as well, a young woman at the verge of a satisfying musical career. To relieve her possible boredom, William made a series of special small telescopes just for her, and eventually she discovered eight new comets. Rather ironically, William's only comet turned out to be a planet! But even before the comets, Caroline began to find new nebulae, and as this biography of the brother-and-sister team shows, her discoveries launched an entirely new tack for William's researches. With his 20-foot instrument with an 18-inch mirror, William began to sweep the sky for nebulae, eventually completing three monumental new catalogues, listing 2500 nebulae.

This brief biography covers in rich detail Herschel's career as the greatest observer of his century and also as a theoretician who constructed explanatory pictures of the structure and evolution of the heavens. Michael Hoskin's long exploration of the history of sidereal astronomy has given him an in-depth context for Herschel's achievement. Today, no one else knows more about the interactions of the William and Caroline team than Michael Hoskin, so it is a very great and secure pleasure to recommend this account, which is both fascinating and authoritative.

Cambridge, October 2012 Owen Gingerich

Contents

Chapter 1
Vocations in Conflict

1.1 The Refugee Musician

In August 1731 the gardener-turned-oboist Isaac Herschel (1707–1767) arrived at Hanover from his family home in central Germany, and enlisted in the band of the Hanoverian Guards. The Elector of Hanover was better known as King George II of Britain, where he resided, but in Hanover he maintained a Court and a regiment of Guards. A year later Isaac hurriedly married Anna Ilse Moritzen (1712/13–1789), an illiterate serving girl from a village outside Hanover. Anna was already pregnant and Sophia, the first of their ten children, was born six months later.

Their fourth child, Friedrich Wilhelm (1738–1822, known to us as William, the name that later became legally his by Act of the British Parliament), was brought up to be a musician, as were all the Herschel boys. In 1753 William joined his father and his older brother Jacob (1734–1792) in the band of the Guards. Soon the French threatened an invasion of Britain, and in 1756 King George summoned the Guards to England to reinforce the army there. The threat soon passed, and the Guards returned to Hanover—but without Jacob, who had little taste for army life and had already secured his discharge.

In 1757 the threat revived, and this time it was against Hanover itself. Isaac and William took part (as non-combatants) in the Battle of Hastenbeck, when the Hanoverians and their allies were defeated by the French. Isaac persuaded himself that William, a boy and not under oath, had no obligation to remain with the army, and so he sent him away to England, along with Jacob, who had been lying low in the family apartment in Hanover while French soldiers occupied much of the rest of the block.

Safely arrived in England, the two refugees survived through their talents as musicians. When peace was restored in 1759 Jacob returned home, but William remained in England. He was technically a deserter, and it would be unwise for him to set foot in Hanover until he received his formal discharge, which he did in 1762 (Fig. 1.1).

M. Hoskin, *William and Caroline Herschel*, SpringerBriefs in Astronomy,
DOI: 10.1007/978-94-007-6875-8_1, © The Author(s) 2014

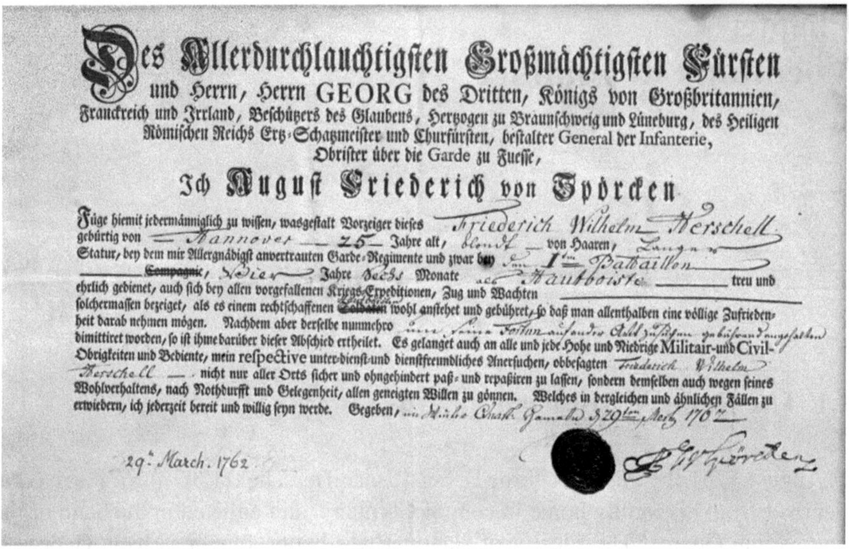

Des Allerdurchlauchtigsten Großmächtigsten Fürsten
und Herrn, Herrn GEORG des Dritten, Königs von Großbritannien,
Franckreich und Irrland, Beschützers des Glaubens, Hertzogen zu Braunschweig und Lüneburg, des Heiligen
Römischen Reichs Ertz-Schatzmeister und Churfürsten, bestalter General der Infanterie,
Obrister über die Garde zu Fuesse,

Ich August Friederich von Spörcken

Füge hiemit jedermänniglich zu wissen, wasgestalt Vorzeiger dieses Friederich Wilhelm Herschell gebürtig von Hannover 25 Jahre alt, blondt von Haaren, Langer Statur, bey dem mir Allergnädigst anvertrauten Garde-Regimente und zwar bey dem 1ten Battaillon ~~Compagnie~~, Vier Jahre Sechs Monate als Hautboiste treu und ehrlich gedienet, auch sich bey allen vorgefallenen Kriegs-Expeditionen, Zug und Wachten solchermassen bezeiget, als es einem rechtschaffenen ~~Soldaten~~ wohl anstehet und gebühret, so daß man allenthalben eine völlige Zufriedenheit darab nehmen mögen. Nachdem aber derselbe nunmehro [illegible] dimittiret worden, so ist ihme darüber dieser Abschied ertheilet. Es gelanget auch an alle und jede Hohe und Niedrige Militair- und Civil-Obrigkeiten und Bediente, mein respective unter-dienst- und dienstfreundliches Ansuchen, obbesagten Friederich Wilhelm Herschell nicht nur aller Orts sicher und ohngehindert paß- und repaßiren zu lassen, sondern demselben auch wegen seines Wohlverhaltens, nach Nothdurfft und Gelegenheit, allen geneigten Willen zu gönnen. Welches in dergleichen und ähnlichen Fällen zu erwiedern, ich jederzeit bereit und willig seyn werde. Gegeben, [illegible] 29ten Martz 1762

29th March. 1762

Fig. 1.1 William's discharge from the Hanoverian Guards. Photograph by the late W. H. Steavenson, courtesy of the Institute of Astronomy, Cambridge University

By then William was making a good if somewhat stressful living as an itinerant musician in the north of England, and he was composing the symphonies and concertos by which he hoped to be remembered. Late in 1766 William (Fig. 1.2) accepted an invitation to Bath in the west of England, to become organist at the Octagon Chapel then under construction. During the season, from the autumn to Easter, Bath was frequented by the aristocracy, and a talented musician could make a good living there. The chapel was a commercial enterprise, where the upper classes would rent pews and then worship their Maker in warmth and comfort. The quality of the music in the chapel was an important selling-point, and William was expected to write music for the choir and train them to cathedral standard. As organist he would receive a regular salary, and this he would supplement by his other musical activities.

In June 1767, some weeks after the death of their father Isaac, Jacob joined William in Bath to help support the family back in Hanover, and he stayed for two profitable years. In July 1770 Jacob paid another (but this time brief) visit to Bath, bringing with him their younger brother Alexander (1745–1821). Alexander was to remain in Bath as a leading cellist until ill-health forced him to retire to Hanover in 1716. His hobby was metalwork, and this would later prove to be of great value to William when he became a builder of telescopes.

Fig. 1.2 William as a young man. This miniature (artist unknown) was probably given by William to Caroline in 1764, when he travelled from England to visit his family in Hanover. Herschel family archives

1.2 The Rescue of Caroline

By 1771 William and Alexander had become increasingly concerned about the fate of their youngest sister Caroline (1750–1848), who was trapped in the family home in Hanover at the beck and call of their domineering mother. Anna enjoyed having her daughter as a household drudge, and so she had prevented her from learning any skills that might enable her to make her escape. A governess of children would be expected to teach French to her changes, and so Anna forbad Caroline to learn the language. She also put obstacles in the way of Caroline's learning the more advanced needlework. Caroline (Fig. 1.3) was fearful for her future, when the brothers now living at home had married and she, a diminutive spinster disfigured by smallpox, was left behind with only her mother for company.

Her brothers contrived an excuse for her rescue by proposing that she come to Bath for a couple of years, to see whether her singing voice was good enough for her to perform as soloist in the Handel oratorios that William regularly promoted. This was most unlikely, for she was a German working-class girl, unused to society, and untrained in any form of music; but it was the only pretext they could invent.

Fig. 1.3 Caroline as a young woman. Such 'shades' could be within reach even of poor families, and exact copies could be cheaply made, either at the time or later. Courtesy of the Museum of the History of Science, Oxford University

In August 1772 William went to Hanover in person to press the matter, and he shrewdly won Anna over by the promise of an annuity to pay for substitute help. Now that Anna was a widow, her eldest son Jacob was head of the household. Caroline, although a grown woman in her twenties, was reluctant to leave without Jacob's formal consent; but, by a happy chance, at the time of William's visit Jacob was away from Hanover with the Court Orchestra, of which he was a member. Letters took a long time to go to and fro, William had pressing engagements back in Bath, and finally Caroline had to make up her own mind. She opted for William and a new life in Bath.

As William and Caroline journeyed across The Netherlands to take ship for England, William introduced his sister to some of the constellations; and when they finally reached London after a terrifying crossing, he took her to opticians' shops. Although she did not realize it until later, William was becoming obsessed with astronomy. As a musician in the north of England he had bought and much enjoyed Professor Robert Smith's book on *Harmonics* (Cambridge, 1749). Recently in Bath he had come across the same author's *A Complete System of Opticks* (Cambridge, 1738), which told the reader a lot about how to make telescopes and a little about what to see with the finished instruments. He had decided to purchase this too, and he had become captivated by its contents.

Caroline accepted that during the Bath season, which was just beginning, William would be much too busy to spare time to train her as a singer. But she was dismayed when the season ended, around Easter 1773, to find that even then he was loathe to devote time to her. He could think of little else but how to acquire telescopes, so that he might view the heavens for himself.

At first William experimented with refractors, in which the incoming starlight passes through a lens. The purpose of the lens is to bring the light to a focus, where the image can be examined through the eyepiece. Unfortunately, the different colours are refracted by the lens through slightly different angles, and this results in a blurring ('chromatic aberration'). The simple answer, as Newton had shown, lay in the reflecting telescope, in which the incoming light passes down the tube and is reflected back by the mirror at the bottom; for all colours are reflected equally. The image near the top of the tube must then be reflected a second time. In 'Newtonian' reflectors, a small flat secondary mirror reflects the image sideways towards the eyepiece, which is inserted in the tube near the top.

William therefore hired a reflector. But whereas almost all observers of his day, both professional and amateur, were preoccupied with the solar system, he had become fascinated with the universe that lay beyond; and to study very distant (and therefore faint) objects, he needed wide-aperture instruments that could collect a large quantity of the incoming light and so bring the objects into view. But large mirrors were expensive to buy, and before long he was grinding his own. This work was boring and each mirror occupied many hours. Caroline found herself reading to him while he worked, and even putting food into his mouth.

The later 1770s saw a clash of wills between the two siblings. To William astronomy was now the first consideration, while Caroline was resolved to pursue her career as a singer. In spite of her brother's preoccupations she persevered, and her career reached its zenith in April 1778, when she sang in Bath as first soloist in *Messiah* (Fig. 1.4). She acquitted herself so well that when the performance ended she was approached and invited to perform in Birmingham, in a concert in which her brother would have no part.

In a fateful decision she chose to decline the offer. But from then on her career as a singer went downhill. William was determined to make time for astronomy, although he was not prepared to sacrifice his income from music. The promotion

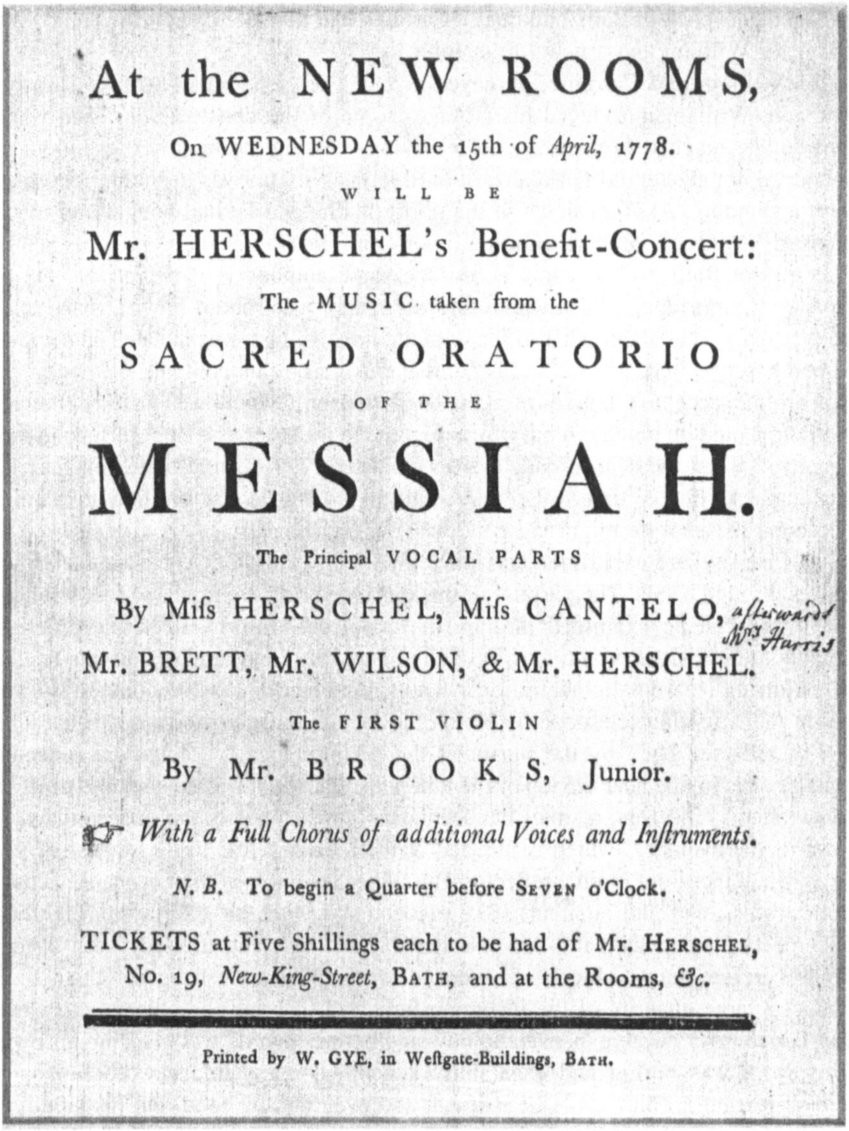

Fig. 1.4 Poster advertising a performance of Handel's *Messiah* in Bath on 15 April 1778, with Caroline as first soloist and William as fifth. Herschel family archives

of oratorios could be profitable; but the rehearsals were time-consuming, and so he reduced them to the minimum. To ensure that the performance was nevertheless of acceptable standard, he took to inviting distinguished soloists from outside Bath, and with them Caroline could not compete.

1.3 The Amateur Astronomer

In March 1774 William had felt sufficiently competent as an amateur astronomer to open his first observing book (Fig. 1.5). He looked at Saturn, and then at the Orion Nebula, best-known of these mysterious milky patches in the sky. There was a dispute among observers as to what nebulae really were. Some thought them vast star clusters disguised by their great distance—clusters so far away that existing telescopes were not able to 'resolve' them into their component stars. Others held that while some nebulae were indeed distant star clusters, others were nearby and formed of a luminous fluid ('true nebulosity' in William's term, 'gaseous' as we would say).

Smith's *Opticks* reprinted a sketch of the Orion Nebula that had in fact been made by Christiaan Huygens as far back as 1656, and William—as yet unaware of how different the same nebula can appear under different observing conditions—decided that it must have changed shape since the sketch was taken. A few decades represented a brief span of time in the context of the large-scale universe. If the nebula had indeed changed so quickly, it could not be a distant and vast star system; it must be small and close at hand, and therefore composed of luminous fluid.

In the years that followed William returned time and again to the Orion Nebula, to test whether or not it had in fact altered shape since last observed. It was not easy to record the appearance of the cloudy region, and so he focused much of his attention on the alignments of the embedded stars. By September 1783 he was convinced that the nebula was "surprisingly changed".

But most of the time the busy musician could spare for his hobby of astronomy he devoted to the building of reflectors. His first observations had been with a 5½-ft reflector—that is, a reflector measuring 5½ feet in the 'focal length' between the primary mirror and the secondary. In May 1776 he made himself a 7-ft Newtonian and then a 10-ft, and all the while he was experimenting with their wooden mountings, for it is of little use to have a fine telescope if you cannot easily point it in the right direction[1] (Fig. 1.6).

In July 1776 he completed a 20-ft with mirrors 12-inches in diameter. In size it was a respectable instrument for an amateur to possess, but it was of limited use because the tube was simply slung from a pole. The plucky observer was perched alongside at the top of an enormous ladder, in the dark, and in peril of life and limb (Fig. 1.7).

In November 1778 William ground and polished for his 7-ft reflector a mirror whose exquisite shape made it quite simply the finest of its kind in existence anywhere. This gave its owner a decisive advantage over all other astronomers, amateur and professional alike: as would become clear in the years that followed, the Bath musician was able to see things hidden from the leading observers of the day. Impressed, his influential admirers would grow in number and in enthusiasm, and eventually they would contrive a means whereby William could abandon music and himself join the ranks of professional astronomers.

---

[1] The classic article on William's telescopes is Bennett (1976).

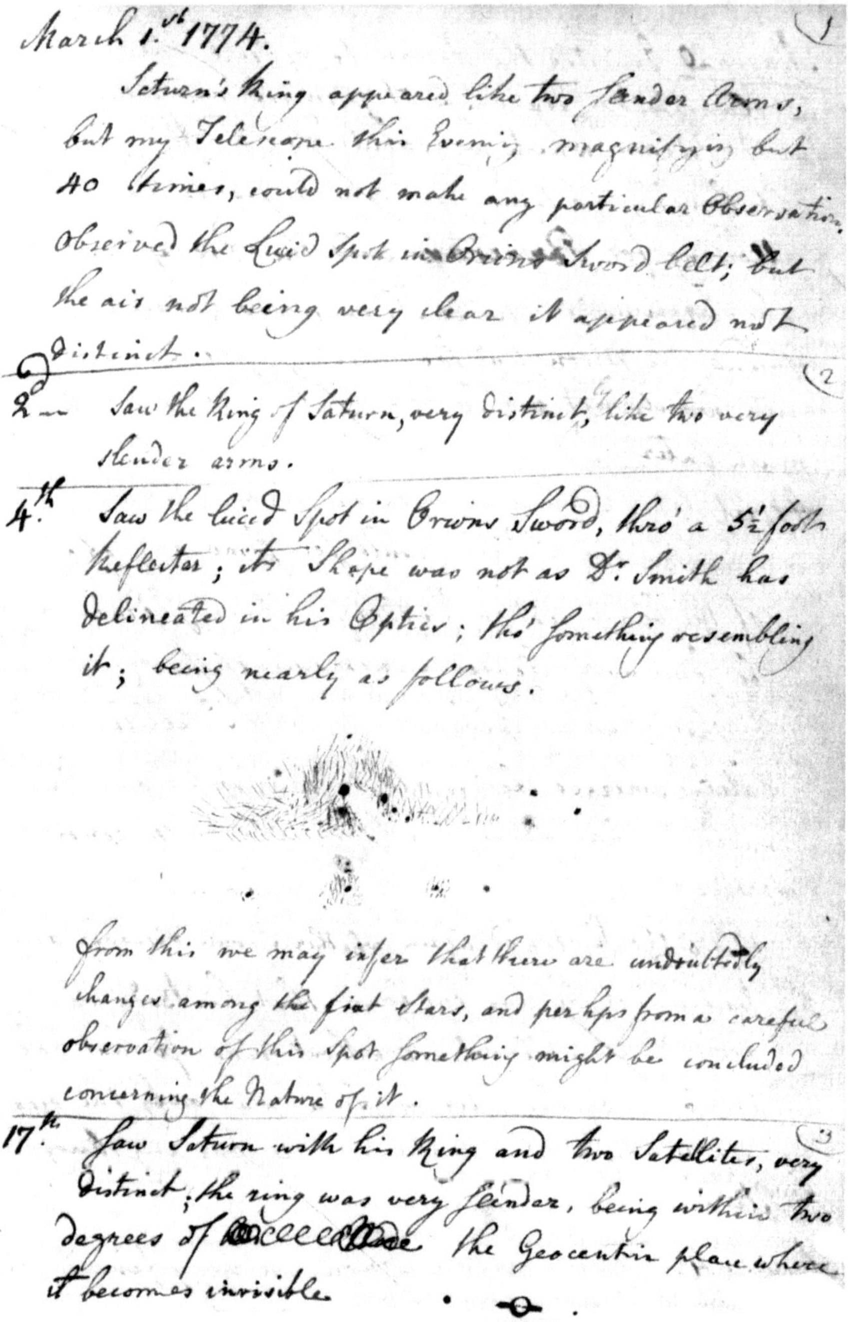

March 1st 1774. (1

Saturn's Ring appeared like two slender Arms, but my Telescope this Evening magnifying but 40 times, could not make any particular Observation. Observed the Lucid Spot in Orion's Sword belt; but the air not being very clear it appeared not distinct.

2 — Saw the Ring of Saturn, very distinct, like two very slender arms. (2

4th Saw the lucid Spot in Orions Sword, thro' a 5½ foot Reflector; its Shape was not as Dr. Smith has delineated in his Optics; tho' something resembling it; being nearly as follows.

From this we may infer that there are undoubtedly changes among the fixt Stars, and perhaps from a careful observation of this Spot something might be concluded concerning the Nature of it.

17th Saw Saturn with his Ring and two Satellites, very distinct; the ring was very slender, being within two degrees of [struck out] the Geocentric place where it becomes invisible. (3

Fig. 1.5 The opening page of William's first observing book. On 4 March 1774 he observed and sketched the Orion Nebula and noted that it seemed to have changed since the time of the sketch by Christiaan Huygens reproduced in Robert Smith's *Opticks*. If so, it could not be a distant and vast star system. Courtesy of the Royal Astronomical Society

Fig. 1.6 A drawing by William Watson of William's 7-ft reflector, which was to become the model for many commercial instruments that he sold in later life. He devised this stand in 1778, and in November of the same year he completed the mirror whose excellence was to give him an advantage over all other astronomers, professional and amateur alike. Courtesy of the Royal Astronomical Society

In his early years as an amateur observer, he had searched for evidence of Lunarians, the intelligent creatures he was convinced populated the Moon; and when they proved elusive he had concentrated on measuring the heights of lunar mountains. But in 1780 he decided to 'review' the brightest stars one by one, so as to get to know them; and when this simple exercise was complete he started over again, this time to examine in turn every one of the naked-eye stars and many more besides.

The motive behind this second 'review' is set out in his notebooks, and it has its origins in the claim made by Copernicus in 1543 that the Earth is in orbit about the Sun. Copernicus's opponents—who understandably found it hard to believe they were passengers on a planet that was racing through space—pointed out that if an observer is moving, then everything he observes will appear to move. If therefore the Earth-based observer is moving from one side of the Sun to the other every six months, each star should appear to move in the opposite direction, in what is termed 'annual parallax'. But even the great Danish observer Tycho Brahe, who at the end of the sixteenth century carried precision observations to the limit of what is possible with the unaided eye, could detect no such movement. By persisting in

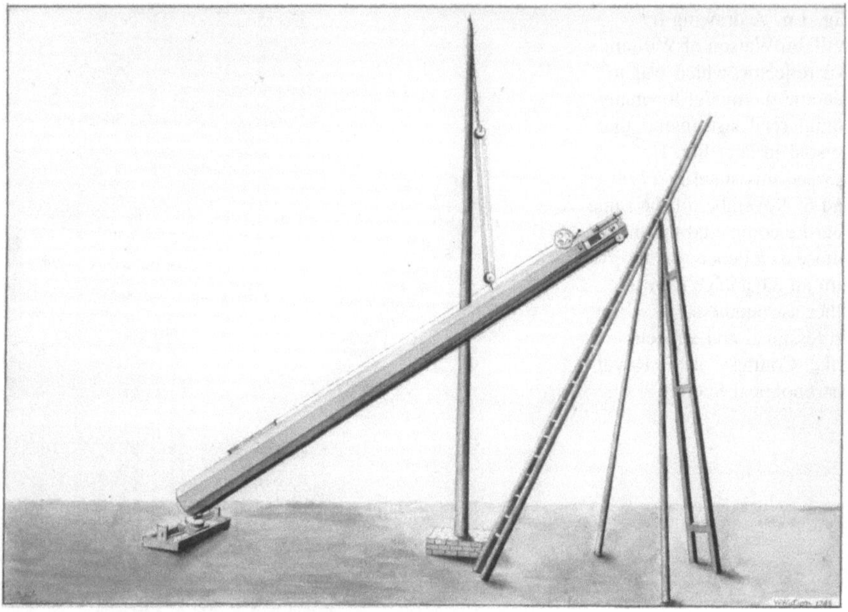

Fig. 1.7 A drawing by William Watson of the 20-ft reflector William constructed in 1776. As in all 'Newtonian' reflectors, the eyepiece is located near the top of the tube. Because the 'focal length' of this reflector—the distance from the primary mirror at the base to the secondary mirror and eyepiece near the top—is 20 feet, this is roughly the height of the observer from the ground when examining sky near the zenith. It must have been nearly impossible for the observer, perched precariously on the ladder in the dark, to make any written records of the observations. Courtesy of the Royal Astronomical Society

their claim, the Copernicans (it seemed) were simply refusing to accept the scientific evidence.

Of course, the further a star from Earth, the smaller its annual parallax will be, and the Copernicans argued (rightly, as it eventually transpired) that the stars were in fact so far away that even Tycho Brahe's instruments were not accurate enough to detect parallax. The whole argument then shifted, as dynamics became part of astronomy: in the later seventeenth century it seemed to many absurd that the massive Sun could be orbiting the tiny Earth. The search for annual parallax then became part of the campaign to determine just how far away the stars are.

Isaac Newton revealed the scale of the challenge by making the plausible assumption that the bright star Sirius is physically the twin of the Sun, and appears less bright than the Sun only because it is farther away. Using an ingenious argument to compare their brightnesses, he showed that the Sun is some million million times brighter than Sirius; and since light falls off with the square of the distance, this implied that Sirius is perhaps a million times further than the Sun.[2]

[2] On Newton's cosmology see Hoskin (1977).

If so, its annual parallax will be minute, no more than the width of a coin at a distance of some miles. Furthermore, this apparent movement of Sirius will take place over a period of six months, during which time the observer's instrument will be subject to the effects of seasonal variations in temperature and humidity. Nor is this the end of the horrors facing the would-be measurer of annual parallax, for changes in atmospheric refraction will also affect the observed position of the star.

Galileo had popularized a way round these difficulties. Suppose two stars lie by chance in virtually the same direction from Earth, so forming a 'double star', and suppose one of the two is very distant while the other is relatively near. The two stars will be equally affected by seasonal variations in the observing instrument, and by changes in atmospheric refraction (and indeed by other problems not as yet identified). But the more distant star will have negligible parallax, and so for all practical purposes will be a fixed point in the sky from which the observer can hope to measure the apparent changes in position of the nearer star. In short, the study of double stars offered the best hope of detecting annual parallax.[3]

All this William sets out lucidly over several pages of his observing book. It is unlikely that he thought this up for himself, and he may well have learned of the double-star method from the Astronomer Royal, Nevil Maskelyne, who had called on the eccentric Bath musician when he was on a visit to the city. The two men had hit it off at once and spent hours in enthusiastic discussion, debating so warmly that Caroline, cloistered in another room of the house, feared they were quarrelling.

Wherever the idea came from, William's second review became a search for double stars, and he had at his disposal his 7-ft reflector with its superb mirror. The very first night he found that the famous Pole Star was a double, and of this more anon.

Until now, William's contact with like-minded students of nature had been chance encounters, but at the end of 1779 all this changed. One night he planned to observe the Moon, but it happened that at the time our satellite was inaccessible from his back garden, and so he took his 7-ft out into the street and observed from there. A passing carriage stopped, and the occupant got out and asked if he too might be allowed a look.

Next day William's new acquaintance called to introduce himself. He was Dr William Watson, a prominent Bath physician and a Fellow of the Royal Society of London, as was his father, a London physician of the same name. Watson Junior became William's devoted friend and ally, and his first step was to invite William to join a society that had been formed only hours earlier, the Bath Philosophical Society, the first provincial scientific society in the land.

William, "optical instrument maker and mathematician", became the Society's most enthusiastic member. At last he was among fellow-spirits. His isolation over, he presented innumerable papers to the meetings of the infant society, which were held in premises a stone's throw from where William was living. His first paper,

[3] For more on the double-star method of detecting annual parallax see Hoskin (1966).

surprisingly enough, was on corallines, and he must have been introduced to natural history by his youngest brother Dietrich (1755–1827), a musician and amateur entomologist who in 1777 had fled the family home in Hanover and ended up spending the next two years with William in Bath.

Dietrich's influence was to prove momentous. Until William appeared on the scene, astronomy had been focused on the study of the individual bodies of the solar system—the Sun, the six planets, their satellites, the occasional comet—each with its own proper name and physical characteristics. The innumerable stars were little more than a backcloth to the motions of the planets; they were for the most part unchanging, and therefore of minor interest.

William the self-taught astronomer was by contrast obsessed with the exploration of the stellar universe, anxious to unravel what he called "the construction of the heavens"; and he would do this by introducing into astronomy the methodology of the natural historian that he had learned from Dietrich. First he would collect double stars by the hundred; later, nebulae and star clusters by the thousand. These specimens he would label and classify, and then consider what lessons might be learned from his collections. It was a whole new way of doing astronomy.

When William completed his second review in August 1781, his collection of double stars had grown to no fewer than 269. As a Fellow of the Royal Society, Watson was entitled to sponsor publication of a paper by a non-Fellow, and he was more than happy to have his name linked with William's list of doubles. It later turned out that a German astronomer had also recently published such a list, but the German's list was modest and there was little duplication.

The most surprising double in William's catalogue was the Pole Star, and his claim that it was two stars rather than one was greeted with raised eyebrows. Even though it is much easier to see what you have been forewarned is there, the English astronomical establishment was quite unable to verify William's claim. Could it be that a Bath musician with a home-made telescope could see things inaccessible to the leading astronomers of the day with their professionally-made instruments?

Astonishingly, it could. At last, in March 1782, a leading English amateur, Alexander Aubert, succeed in confirming that the Pole Star is indeed double. Sir Joseph Banks, President of the Royal Society, wrote personally to William to offer his congratulations. This success was proof that a telescope-builder and observer of extraordinary talent—and perseverance—had appeared out of nowhere. Not only that, but William had already shown himself to be a bold theorist: in 1781 he had published a paper with the Royal Society suggesting a method by which future generations could determine whether or not the daily rotation of the Earth is subject to change, a possibility that few had previously considered. Yet this remarkable man was currently being forced to spend—waste—most of his time earning his living by teaching and performing music. Something must be done about this; but what?

William's ambitions meanwhile were growing. His double stars were bright objects, and in cosmological terms close at hand; but what of the faint and mysterious nebulae, whose physical nature was the great unsolved-mystery of what we would call deep-sky astronomy? So far his biggest mirror was a mere twelve inches in diameter, and something much larger would be required to collect light enough to bring nebulae into view in numbers. And so, at the beginning of 1781, he resolved to build himself a monster telescope, 30 feet in length and with mirrors 3 feet in diameter (Fig. 1.8).

He spent the early months of the year in planning the mounting that was to carry such a heavy tube, and in experimenting with the materials best suited for the metal mirror. Instead of settling for the usual alloy of copper and tin he decided to investigate the possibilities of iron, encouraged by the gloss achieved by makers of buckles and candle snuffers. But at a Bath foundry the iron took no less than eight hours to turn molten, and even then the small experimental disk proved to be full of holes. Unfazed by this, William took the work to an experienced iron-founder in nearby Bristol, and then spent hours polishing the resulting mirror; but in the end he had to admit it was not reflective enough to be of use to him.

Eventually William was forced back to the traditional copper and tin. But then it turned out that there was not a single metal-founder in either Bath or nearby Bristol who could cast a disk with diameter as large as 3 feet. Undeterred, William decided to build a furnace in the basement of his own home, and to take charge of the casting himself. The mould was to be of horse dung, in the medieval tradition of the casters of church bells, and family and visitors alike—even the eminent Dr William Watson, FRS—were put to work to pound the vast quantity needed.

Alas, on the first attempt the mirror cracked in cooling; and on the second the mould broke and the molten metal ran out onto the flagstones, whose scarring can be seen to this day. Some of the stones exploded with the intense heat. Caroline, who had been banished to safety at the bottom of the garden, was alarmed to see William and Alexander burst out of one door as they fled for their lives, while the founder and his assistants ran in the other direction. At this even William decided that enough was enough—for the present.[4]

William and Caroline had moved into the house in question—19 New King Street, now the home of the Herschel Museum—early in March. More exactly, William had moved there, leaving Caroline for a week or two in their previous home. This was because William had made an ill-judged investment in a millinery business, which was located on their ground floor. The site was unsuitable, the business had failed, and the stock was being sold off. Caroline was to remain behind for a while to make sure they were not being cheated. And so it was that she was elsewhere when an event took place in New King Street that would change both their lives.

---

[4] William describes his attempts to build a 30-ft reflector in Royal Astronomical Society [RAS] Herschel Archive W.5/12.1, 48–58.

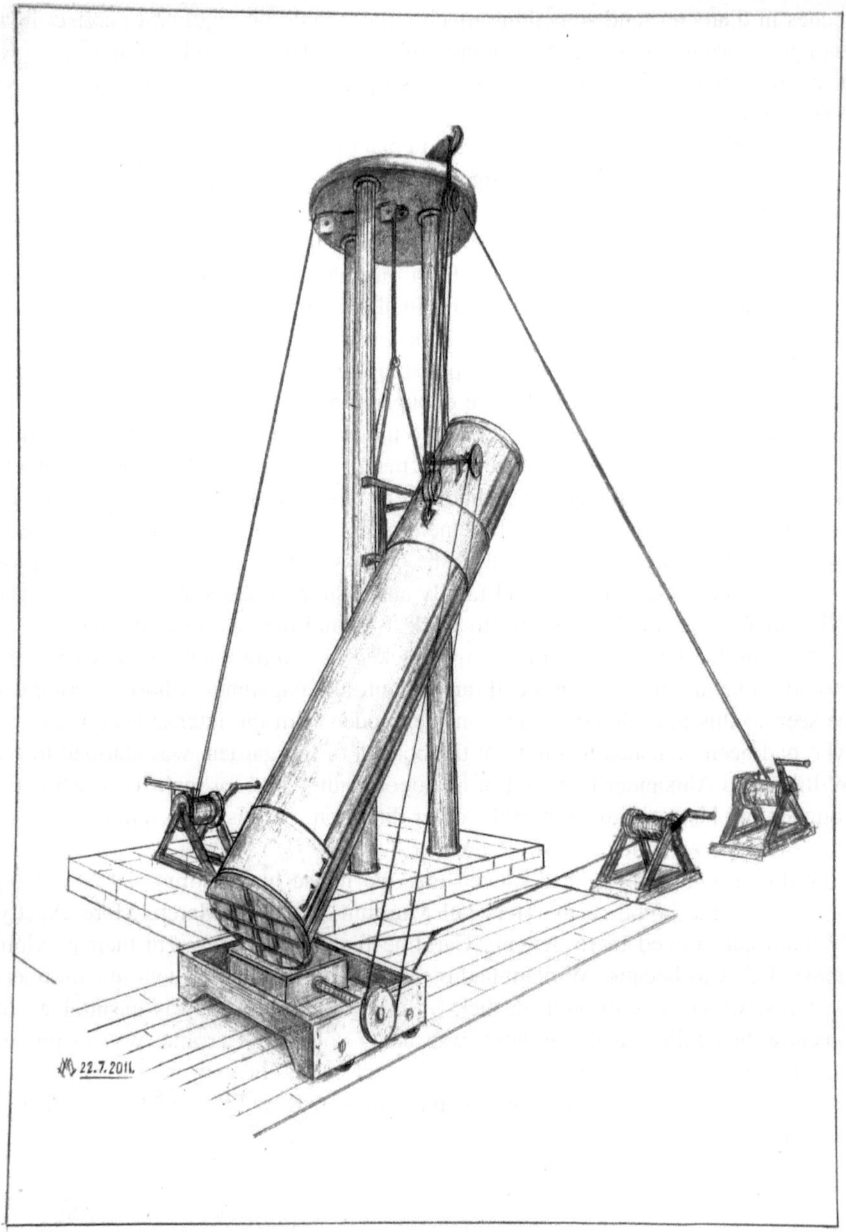

Fig. 1.8 A reconstruction by Andreas Maurer, based on the accounts in William's notebooks, of what the 30-ft reflector William attempted to make in 1781 would have looked like if completed

1.4 A New Planet

On the evening of 13 March 1781, William's review with his matchless reflector was taking him as usual from one star to the next, when he came to one object that he could see at a glance was no ordinary star. His curiosity aroused, he returned to it four days later, and he found that it had already altered position relative to the nearby stars. An object that moved perceptibly in a mere four days must be close at hand, within the solar system. William, knowing nothing of the current speculations concerning an undiscovered planet in the puzzlingly-large gap between Mars and Jupiter, assumed it must be a comet, although it did not have the appearance of a typical comet. And so he prepared for the Bath Philosophical Society a paper entitled "Account of a Comet". Watson forwarded the paper to the Royal Society in London, and in due course it appeared in the Society's *Philosophical Transactions*.

Comets then as now were of lively interest to astronomers and the public alike, and at Greenwich Observatory Nevil Maskelyne searched in the area of sky that William had specified. But in the Greenwich telescopes every object thereabouts looked to be an ordinary star, and it took Maskelyne some time to identify one that was moving. At Oxford Professor Thomas Hornsby had no more success, and finally was driven to write to William and ask for a sketch of the sky with the 'comet' marked. There could be no better proof of the superlative quality of the mirror that William had ground and polished for his 7-ft reflector.

Hornsby continued to believe for months that the object was indeed a comet, but Maskelyne was not so sure. It might be a planet, and if so, it would be the first to be discovered since the dawn of history. And when the mathematicians had enough data to compute an orbit, Maskelyne was shown to be right: an unknown Bath musician, whose name few could spell correctly, had doubled the scale of the known planetary system. In November 1781 William was awarded the prestigious Copley Medal of the Royal Society, and the following month he was elected Fellow.

William's allies could hardly believe their luck. For centuries the customs of royal and aristocratic patronage had prescribed that, if a king or nobleman accepted the dedication of a literary or artistic work, he was then under an obligation to reward the author or artist, usually in financial terms. Although such patronage was now rare, it was not unknown in the court of George III. Indeed, William's elder brother Jacob had exploited the customs in the late 1760s, by dedicating a set of sonatas to Queen Charlotte. Jacob was summoned to court, and thus the King and Queen encountered a Herschel for the first time. Jacob's performance there was so impressive that the King gave instructions for his salary as a member of the Hanoverian Court Orchestra to be increased.

It was of course very rare for an observational astronomer to have the opportunity to offer a dedication, but there was a famous precedent. In 1610 Galileo had dedicated his discovery of four moons of Jupiter to Duke Cosimo, whose Medici family ruled Florence, and he named them "the Medicean Stars"; and in return he had been appointed mathematician and philosopher to the ducal court. Sir Joseph

Banks, the President of the Royal Society, was as shrewd an operator as you could meet, and he had the ear of the King. Banks had no doubt that William should dedicate his planet to George. The King would surely be thrilled at the prospect that for all future time the Sun's family of planets would be known as Mercury, Venus, Earth, Mars, Jupiter, Saturn, George.

The King may have had his share of vanity, but he was also an intelligent and enlightened man, and a serious student of astronomy. Indeed he had built an observatory next to Kew Palace, to the west of London, explicitly in order that he himself might observe in 1769 the very rare transit of Venus across the face of the Sun; and he had appointed his former tutor, Charles Demainbray, as observer there. In the English climate, to go to such lengths to observe a once-only event was tempting fate; but luck was with him, and on the appointed day the King and Demainbray were able to watch the tiny dot as it crossed the face of the Sun.

When William discovered his planet, Demainbray was in his seventies and in declining health, and his post at Kew had become a sinecure. He did little more than make the observations he needed to correct the official clocks in the neighbourhood of Westminster. And so, on 23 February 1782, Banks attended the royal levee with the intention of suggesting to the King that he nominate William to succeed Demainbray when the time came.

To Banks's dismay, what was to have been a leisurely exercise in gentle diplomacy was thrown into confusion by the news that Demainbray had just died. Banks—and Watson—hoped that the King would forthwith nominate William as the royal observer at Kew. The King had heard a lot about William's prodigious talent in astronomy, and he knew how gifted were other Herschels of Hanover—he had heard Jacob perform, and two of Sophia's sons were musician's in Queen Charlotte's band—so the prospects were good. But for some reason the King declined to go ahead with the proposed nomination. Weeks later it transpired that he had already promised the post to Demainbray's son, and he was unwilling to go back on his word.

Banks and his allies pursued their plan of having William name his planet for the King, and thus put the King under an obligation. For this the King must of course indicate his consent in advance, and the conspirators selected Colonel John Walsh of the Worcester Regiment to put the matter to him face to face. This Walsh did at the end of April. As he told William, "I took occasion to mention that You had a twofold claim, as a native of Hanover and a Resident of Great Britain, where the Discovery was made, to be permitted to name the Planet from His Majesty".[5]

The King was now in a quandary. He would be glad to foster astronomy by enabling William to give up music and dedicate his life to science, and he would be more than glad to have the planet named George. But how could such a change of career be brought about? Kew aside, there was no obvious way forward. And so the King bought himself time, by summoning William to London, along with his 7-ft reflector.

[5] John Walsh to W. Herschel, 10 May 1782, RAS Herschel Archive W.1/13.W.5.

In these circumstances William was finding it almost impossible to concentrate on music. One morning he was about to set off with Caroline for Bristol, where he was to direct a Handel oratorio that evening, and they were taking with them the scores needed by the orchestra and choir. William was deep in conversation with William Watson about his prospects of becoming a professional astronomer, when one of Sophia's sons arrived from Windsor with confirmation that William was to present himself at court. With her brother's mind in turmoil, Caroline did what she could to assemble the necessary music, and off they set for Bristol. Not surprisingly, the concert was a shambles, as eloquent complaints in the local paper testify. A chastened William made sure that nothing of the sort would happen again, by hiring only soloists and choirs of the highest quality.

The stand of William's 7-ft could not easily be transported by coach, but the essential components were the optics, and William devised a substitute stand that he could disassemble for the journey to London and reassemble on arrival. He did not know whether the King was a serious observer, but he wisely worked on this assumption, preparing a short but demanding list of objects to impress the royal eyes.

The King was still at a loss as to what to do for William, and so he bought more time by instructing him to take his reflector to Greenwich so that it could be appraised by the Astronomer Royal. Trials duly took place, and William was delighted by the outcome.

> These last two nights I have been star-gazing at Greenwich with Dr Maskelyne and Mr Aubert. We have compared our telescopes together, and mine was found very superior to any of the Royal Observatory. Double stars they could not see with their instruments I had the pleasure to show them very plainly, and my mechanism is so much approved of that Dr Maskelyne has already ordered a model to be taken from mine and a stand made by it to his reflector. He is however now so much out of love with his instrument that he begins to doubt whether it *deserves* a new stand.[6]

William was still more delighted after he took his 7-ft to the home of Aubert, who had used his wealth to assemble an outstanding collection of astronomical instruments. "I can now say", William wrote to Alexander, "that I absolutely have the best telescopes that were ever made",[7] and this was no more than the truth.

Weeks passed, and Caroline and Alexander began to run out of excuses for William's absence from Bath and his resulting inability to give his scheduled musical lessons. The suggestion was floated that he might become King's astronomer in Hanover, a post to be created for him; but the proposed salary was a mere £100, less than a quarter what William was earning from music. And then an idea occurred to the King. He would be glad to have his own astronomer-in-residence, to be on call when his mood took him. Not only that, but this would solve the problem of how to entertain dinner guests at Windsor Castle after the meal was finished. If William's observatory were only a few minutes' ride away,

---

[6] W. Herschel to C. Herschel, 3 June 1782, Lubbock (1933), 115.

[7] W. Herschel to A. Herschel, 10 June 1782, Lubbock (1933), 116–117.

the guests could be despatched to look through his telescopes—and to admire the King's patronage of science.

The question was whether William had the charm and diplomacy to move comfortably in the highest ranks of society, and so a final trial was arranged, to be held at Windsor Castle, and this time he was to demonstrate the heavens to the Royal Family themselves. This trial too was a complete success, for William was a man of innate courtesy. And so a deal was struck. William would quit Bath and move to the neighbourhood of Windsor, where he would devote himself to astronomy. His only duties would be, on request, either to bring a portable telescope to the castle or to receive royal guests in his own observatory. In return he was to receive a 'pension' of £200, a typical salary of a middle-ranking Court official.

Chapter 2
The Construction of the Heavens

2.1 The King's Astronomer

The modest size of the pension outraged William Watson, but he was routinely outraged on William's behalf. For his part, William could not abide the thought of returning to the drudgery of forty hours of musical instruction every week, and so he accepted the royal offer with alacrity, even though his future income would be less than half his current earnings. He took it for granted that Caroline would willingly abandon her career in music, and would quit Bath, one of the cultural centres of the land, in favour of Datchet, a tiny village a couple of miles east of Windsor Castle where nothing ever happened. He would have to find some way for her to occupy her time, and he decided the answer was for her to become his assistant astronomer. But it would take Caroline a while to realize that this was the fate in store for her.

Back in Bath, William hired a wagon, and for some days it stood outside their home in New King Street while their furniture was loaded. The pole that supported the 20-ft must have presented a challenge. Finally all was ready, and early in the morning of 30 July 1782 the wagon set off for Datchet. William had rented a rambling property which, though dilapidated, had ample space to meet his needs as an astronomer and builder of telescopes. William himself took the morning coach for London and alighted at the crossroads in the village of Slough, a mile or so due north of Windsor. There he took a room for the night in one of the coaching inns. Perhaps he chose the Crown, whose landlady was a Mrs Elizabeth Baldwin; she would one day be his mother-in-law.

The following evening Caroline joined him, along with Alexander who was to help them settle into their new home. The three of them walked over to Datchet where they took rooms in the Five Bells Inn (now the Royal Stag), next to the church. The following morning they awoke to find that the wagon had safely arrived and unloading could begin.

Caroline now had her first sight of her new home, and she was aghast. It was "the ruins of a place" and had not been inhabited for years. Even when they eventually left the property three years later "there was not one room but where

M. Hoskin, *William and Caroline Herschel*, SpringerBriefs in Astronomy,
DOI: 10.1007/978-94-007-6875-8_2,

the rain came through the ceeling",[1] and perhaps it was fortunate that royal visits would always take place in fine weather.

There was no sign of the servant that William had engaged, and it transpired that she was in prison for theft. As for William's optimistic forecast that the cost of living would be cheaper in the country than it had been in Bath, this proved not to be the case, and Caroline particularly resented the exorbitant prices charged by the butcher in the village.

Alexander spent some time helping them to settle into their new home, and William and he got the 20-ft reflector erected. The garden was little short of a jungle, and Alexander narrowly escaped falling down a well. After this, men were called into mow the grass, and only then could the brothers see where they could safely set foot.

Before long it was time for Alexander to return to Bath for the start of the season, and the silence that then descended on the house brought home to Caroline the transformation in her life, and the loneliness and isolation that was now to be her lot. William by contrast was exhilarated. He was a professional astronomer at last—but he was already in middle age, and there was so much for him to do. He was currently engaged in a third 'review', and on only their second night at Datchet he was already at work with the 7-ft, examining stars as faint as twelfth magnitude in his continuing quest for doubles. This review would result in a second catalogue, this time of 434 specimens.

There was a downside. The King was much enjoying having an astronomer on call, and from time to time he summoned William to the castle with his portable reflector. By the time the royal whims had been satisfied and William had returned to Datchet, valuable observing time had been lost, to say nothing of the expense involved in transporting the telescope. Fortunately a break in the routine occurred when William went to Bath to collect the valuable fragments of the 3-ft mirror. He and Alexander took the opportunity to cast a 12-inch mirror for the 20-ft, and it was well that they did so for at the New Year the existing mirror shattered with the frost.

King George was keen to see the 20-ft for himself, but for this he had to visit the Herschel home. Once this precedent had been established, William's summonses to the Castle became less frequent.

2.2 Stars No Longer Fixed

In the decades that lay ahead, William's many papers in *Philosophical Transactions* almost always exploited his own unrivalled access to the observational evidence, but soon after his move to Datchet he undertook an investigation that

[1] Hoskin (2003d), 79.

depended entirely on desk-work and used data available to all: the determination of the direction in which the Sun and the solar system are moving through space.[2] The Greeks had contrasted the 'wandering stars' or planets, which had individual movements across the sky, with the 'fixed stars', all of which (it seemed) preserved for ever their positions relative to their neighbours. But in the mid-seventeenth century René Descartes had shown that the fixed stars were simply distant suns and the Sun our nearby fixed star. Later in the century Isaac Newton had argued in his *Principia* that all matter in the universe pulled all other matter in accordance with the law of gravity—in which case every star pulled every other star. Why then did they remained fixed?

At last, in 1718, Edmond Halley found that three stars were no longer in the positions assigned to them in Antiquity—they had 'proper motions' of their own. But if stars were changing their location, the gravitational pull they exerted on our Sun would also change, and the Sun too would begin to move (even if it were not doing so already). The solar system therefore must be in motion; but how could one determine the direction in which it is heading?

When we look at the night sky from Earth, we see that every one of the heavenly bodies appears to rotate overhead, in movements that fall into a pattern. We decide that it is preferable to ascribe the pattern to ourselves rather than to an inexplicable concidence in the movements of the innumerable stars, and so we infer from the pattern that we are observing the heavens from an Earth that is rotating.

The pattern of movements that will result from a displacement of the Sun and solar system is equally clear. If we walk towards a forest, the trees to the left of the direction in which we are heading appear to move further away to the left, and the trees to the right of the direction, further away to the right. By analogy, if the Sun is moving towards a given point in the sky (which William would call 'the solar apex'), the stars will appear to move away from that point. The challenge, then, was to identify such a pattern in the known proper motions of stars.

A star with a measurable proper motion will appear to travel steadily across the sky, year on year. Its annual (angular) movement is therefore the change between its position at some date in the past and its position now, divided by the number of years that have elapsed. The problem in William's day was that two troublesome factors that affected the observed position of a given star—aberration and nutation—had only recently been recognized. As a result, a star's position around the turn of the century as given in John Flamsteed's great British Catalogue of stars was subject to errors unsuspected by Flamsteed, and so the listed position provided an inaccurate starting point for determination of the star's proper motion. The passage of the years would gradually improve the situation; but for the present the number of reliable proper motions was small.

They did however include the handful listed by the meticulous Nevil Maskelyne in his 1776 volume of Greenwich observations: seven motions in the

[2] For details of this investigation see Hoskin (1966).

component of right ascension (the equivalent of longitude), along with two in declination (the equivalent of latitude). William found that the directions of every one of the seven motions in right ascension could be explained as the natural consequence of the Sun's travelling towards a region of sky that included the constellation Hercules.

William came across a second list of (supposed) proper motions in the multi-volume *Astronomie* of Jerôme Lalande, the Paris astronomer who was to become one of his (and Caroline's) greatest admirers. Lalande had drawn on a list of no fewer than eighty purported proper motions compiled by the Göttingen observer Tobias Mayer in 1760. Mayer had accepted that many of these 'motions' would be spurious, the result of observational error; but even when this was taken into account, he was quite unable to perceive any pattern in his data.

Lalande decided to include in his book only those twelve Mayer 'motions' that corresponded to a change of position of at least 18″ (in one or other coordinate) that had taken place over the half-century or so in question. These changes, he thought, were large enough to be considered genuine. For the sake of completeness he also listed the nominal motion of the star in the other coordinate, even if this was negligible and should properly be disregarded. Thus the bright star Aldebaran was listed as having moved a plausible 18″ in declination and a token 3″ in right ascension.

Unfortunately William had a simplistic attitude to scientific method: for him, facts were facts and theories were theories. Faced with Lalande's list, he acted as though all the angles listed, large and small, were equally trustworthy. Happily, he found that his explanation of the seven Maskelyne motions in right ascension would work equally well for the great majority of Lalande's motions. Furthermore, if he took at face value the 3″ 'motion' in right ascension of Aldebaran, he could halve the region of sky he had derived from Maskelyne.

He then looked into the motions in declination, and he concluded that by assigning the Sun an apex close to the star λ Herculis, he could account for the overwhelming majority of these Maskelyne/Lalande motions (although not for quite as many as he at first thought, for the geometry involved is not simple). This proposed apex can strike the modern astronomer as nothing short of amazing, for almost the same apex is accepted today; but William's improper use of the 3″ had been a lucky feature of his reasoning. All the same, there is no denying that William demonstrated remarkable originality and boldness in the paper that he sent to *Philosophical Transactions* just six months after he turned professional.[3]

Never one to let well alone, William returned to the same theme two decades later, in papers published in 1805 and 1806.[4] This time he attempted not only to

---

[3] W. Herschel, On the proper motion of the Sun and solar system, *Phil. Trans.*, 73 (1783), 247–283, reprinted in Hoskin (2012).

[4] W. Herschel, On the direction and velocity of the motion of the Sun, and solar system, *Phil. Trans.*, 95 (1805), 233–256, and On the quantity and velocity of the solar motion, *Phil. Trans.*, 96 (1806), 205–237.

revise the position of the apex in the light of new data but also to make an estimate of the speed of the solar motion. Over the decades the ever-reliable Nevil Maskelyne had been continuing his observations at Greenwich, and he now had data on the proper motions of 36 stars. These William plotted on a celestial globe. But William's simple approach could not easily cope with so much information, and he found himself mathematically out of his depth when attempting to establish a revised apex.

As to the speed of the solar motion, this could be estimated only from an analysis of the speeds with which stars appear to move relative to ourselves. These speeds have two components: radial (the speeds with which the stars approach or recede from the observer) and transverse (at right-angles to the line of sight). In William's day astronomers had no information whatsoever about the radial velocities of stars. The transverse velocities could in principle be determined: from the star's proper motion—its angular speed across the sky—multiplied by its distance. The problem here was that its distance could be estimated only on the (wholly erroneous) assumption that all stars are physically similar. As a result both of William's papers are unsatisfactory.

In thus seeking to understand changes in the observed positions of the 'fixed' stars, William was bold and innovative; but when it came to changes in their apparent brightnesses he was conservative. Early modern astronomers had first been alerted to such a change in 1572, when a supernova ('Tycho's nova') shone forth, so bright that it could be seen in the daytime. The youthful Tycho Brahe had been able to show that the apparition was celestial, rather than atmospheric as Aristotelian philosophy insisted.

Other new stars appeared and disappeared around the turn of the seventeenth century. Then, in 1638, a star that became known as Mira Ceti ("the wonderful star in the Whale") appeared, disappeared—and reappeared again. By 1667 Ismaël Boulliau noticed that although this star was not always of the same brightness at maximum, these maxima were occurring at regular intervals, of some 333 days; and to this extent Mira was lawlike in its behaviour. Boulliau suggested that the star might be rotating, as does the Sun—but that whereas the Sun has only small dark spots (which were known to vary), Mira might have large dark patches (and these might likewise vary). The cyclic behaviour of Mira would then be the result of its rotation every 333 days, its non-cyclic behaviour the result of variations in the dark patches.

For over a century thereafter there were few notable advances in the study of variable stars. The topic had fallen into disrepute because the 'dark patches' explanation was flexible and could, in the Greek phrase, 'save the phenomena' all-too-readily; and claims to have detected yet another variable star were easy to make and impossible to challenge.

Interest was revived by two English amateurs,[5] one of whom—Edward Pigott—had visited William in Bath in 1778 and observed with him. Pigott's father was a

[5] For more on Goodricke and Pigott see Hoskin (1979).

keen amateur observer of the old school, who when the family moved to York in 1780 set about building himself "a magnificent observatory" with conventional fixed instruments. Edward preferred to look for different territories to conquer, and towards the end of 1781 he decided to re-examine the seventeenth-century accounts of variable stars.

Near the Pigott home lived the seventeen-year-old John Goodricke, a deaf-mute. He and Edward Pigott became friends, Pigott introduced Goodricke to variable stars, and in the summer of 1782 they both began to keep watch on stars reputed to be variable. These included Mira, and also Algol, or β Persei, a second-magnitude star whose variability had been reported by Geminiano Montanari in 1667. On 12 November Goodricke was astonished to find Algol was only of fourth magnitude, but when both men examined it the following night it was back to second. They kept watch on the star in the weeks that followed, but it was always of its usual brightness. Then, on the night of 28 December, both men at first found it to be fourth magnitude, and then saw it brighten to second magnitude before their very eyes. Nothing like this had ever been seen before.

The next day Pigott sent Goodricke a note with the novel suggestion that Algol had a satellite that was eclipsing its parent star at regular intervals. The two accordingly set out to keep careful watch on Algol to determine the period of the supposed satellite's orbit, and finally arrived at the astonishing value of under three days.

Pigott now sent news of these developments to Maskelyne, asking him to pass word to William. Maskelyne's letter reached William on 27 April 1783, and he at once began to keep watch on the star. Three days later Banks also wrote to William with the news, and the same evening he came to Datchet in the hope of seeing the drop in brightness for himself—but without success, for they had made a minor error in calculating the time it would occur.

On 8 May William prepared to go to London for the meeting of the Royal Society, and he put his own observations of Algol in his pocket. The phenomenon was apparently public knowledge, and he had no reason not to share his data with the other Fellows. Only later did he discover that Pigott had departed York a month earlier, leaving the publication of their results in Goodricke's hands; the teenage deafmute, unused to such matters, had delayed submission, and as a result William had inadvertently stolen his thunder.

In the paper, the cautious Goodricke had offered the traditional 'dark patches' explanation alongside the novel proposal that Algol was being regularly eclipsed. As theories the two were very different: the supposed 'dark patches' were flexible enough to account for almost any changes and so could not easily be disproved; by contrast the eclipse hypothesis made precise predictions, that the 'light curve' of Algol would fall and rise with perfect symmetry, and that this would happen with perfect regularity. It may be that the two friends were deceived (by changes in the 'seeing') into thinking that these predictions had not been fulfilled; or their confidence in the eclipse hypothesis may have been shaken when they found that three other short-period variables that they discovered soon after could not be explained in this way. Whatever their reason, they abandoned the hypothesis that we now know to be correct.

For his part, William was uncharacteristically unadventurous in his approach to variable stars. As he wrote in 1796,

> Dark spots, or large portions of the surface, less luminous than the rest, turned alternately in certain directions, either towards or from us, will account for all the phaenomena of periodical changes in the lustre of the stars, so satisfactorily, that we certainly need not look out for any other cause.[6]

But he did make a massive contribution to the century-old problem of how to confirm that a given star had in fact changed in brightness. Pigott and Goodricke had supplied the clue when they adopted the practice of listing the stars in a given region of sky in strict order of brightness. If one of the stars was variable, it would betray this by disturbing the listed order from time to time. William adopted this technique but on a massive scale, by publishing in the late 1790s four "catalogues of the comparative brightness of the stars". The Harvard astronomer E. C. Pickering was to comment a century later: "Herschel furnished observations of nearly 3,000 stars, from which their magnitudes a hundred years ago can now be determined with an accuracy approaching that of the best modern catalogues."[7]

2.3 Sweeping for Nebulae

Caroline meanwhile was pining for the social and musical life she had enjoyed so much in Bath, and to help her to adjust to their new lifestyle William contrived for her a little refracting telescope with which she was to 'sweep' the heavens. The tube was positioned against a vertical spindle, and Caroline was to keep the elevation of the tube unchanged while she rotated it around the spindle, thus searching or 'sweeping' a horizontal strip of sky. This done, she should alter the elevation a little, and do the same again. In this way she was to search a region of sky, looking out for anything of interest—double stars, comets, nebulae, whatever. At first she felt lonely and chilled while sweeping, outside in the dark and sometimes without a human being within call; but after she came across her first nebulae she began to warm to the task William had assigned her.

William was able to confirm that these early nebulae of Caroline's were in the list of nebulae and clusters compiled by the French observer Charles Messier. Messier was a renowned comet-hunter, and sometimes he had found himself wasting valuable time investigating a milky patch in the sky that looked like a newly-arriving comet, but which proved to be a permanent nebula and so of no interest to him. He had therefore compiled over the years a catalogue of nebulae,

---

[6] W. Herschel, On the periodical star α Herculis, *Phil. Trans.*, 86 (1796), 452–482, p. 455.

[7] E. C. Pickering, *Harvard Annals*, 23 (1890), 231.

and this reached its final form in 1781. We still use the catalogue today: M42, the Orion Nebula, is no. 42 in Messier's catalogue.

In December 1781 William Watson had come across a copy of what was in fact the second version of the catalogue, comprising seventy objects. He had bought it and sent it as a present to William; and it was in this second version that Caroline's early nebulae were to be found. On the night of 26 February 1783, however, Caroline came across a nebula of which she could proudly write: "Messier has it not." Caroline, the novice observer with a telescope that was little more than a toy, had (it seemed) discovered a nebula unknown to science (Fig. 2.1).

In fact the nebula was M93, no. 93 in Messier's final catalogue of 103 objects. But William and Caroline were unaware of the existence of this final version, and it seemed to them that Caroline had made a notable discovery. Nor was this all, for later the same night Caroline was again writing "Messier has it not"; and this time her nebula was indeed unknown to the Frenchman.

William was mightily impressed. Evidently there was a rich harvest to be gathered among the nebulae. Impulsively, he interrupted the routine of his search for double stars and began himself to 'sweep' for nebulae, with a modest refractor. But he soon realized his folly: these small instruments with wide fields of view were appropriate when one was searching for a comet that perhaps had not been there to be seen the previous night; but nebulae were part of the permanent night sky, and should be examined carefully and unhurriedly, with the largest telescope available.

Back in August 1782, as soon as he had settled into their Datchet home, William had begun to plan a new telescope—not as large as the over-ambitious instrument he had attempted to build in Bath, but a substantial improvement on the existing 20-ft. To pay for it he drew on the savings he had accumulated while in Bath. The reflector came into service in the autumn of 1783, a very convenient timing as William had just completed his search for double stars (Fig. 2.2).

The new reflector was again a 20-ft, but it had 18-inch mirrors and, more importantly, a stable ladder-type construction. In its early months it was fixed facing towards the south. If the observer intended to make extended observations of individual objects, he would install the impressive observing gallery on which to stand. He could then, by brute force, drag the tube a little to one side, find the object he wanted, and follow it as it transited the meridian. If however he intended simply to allow the rotation of the sky to bring objects into view as their turn came, he might choose to observe from a chair mounted on the ladderwork.

During the summer of 1783 William had taken time out to familiarize himself with some of Messier's nebulae, using whichever of his existing instruments came most easily to hand. Now, in the closing weeks of 1783, he used the new, 'large' 20-ft in solo attempts to sweep for nebulae. Its predecessor, the 'small' 20-ft, now stood idle, something that William could not abide; and so he assigned it to Caroline, and told her to re-examine various of his double stars. She, poor thing,

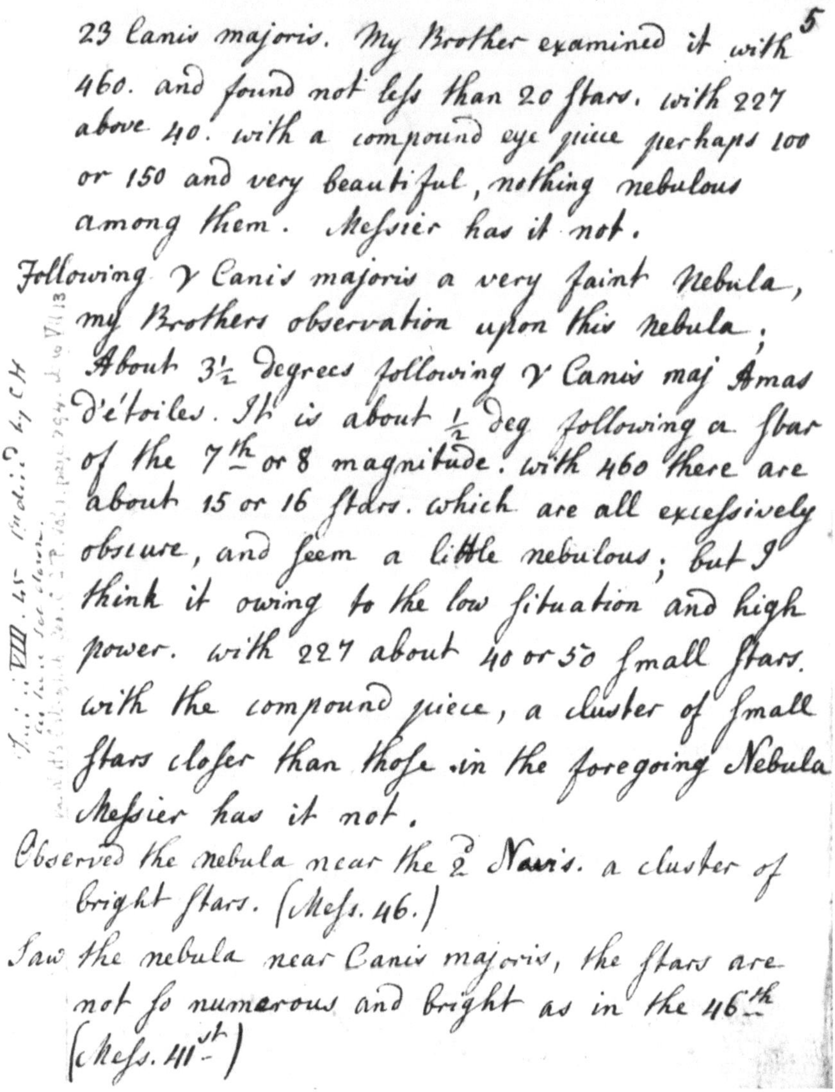
23 Canis majoris. My Brother examined it with 5
460. and found not leſs than 20 ſtars. with 227
above 40. with a compound eye piece perhaps 100
or 150 and very beautiful, nothing nebulous
among them. Meſsier has it not.
Following γ Canis majoris a very faint Nebula,
my Brothers observation upon this nebula;
About 3½ degrees following γ Canis maj Amas
d'étoiles. It is about ½ deg following a ſtar
of the 7th or 8 magnitude. with 460 there are
about 15 or 16 ſtars. which are all exceſsively
obscure, and ſeem a little nebulous; but I
think it owing to the low ſituation and high
power. with 227 about 40 or 50 ſmall ſtars.
with the compound piece, a cluster of ſmall
ſtars closer than thoſe in the foregoing Nebula
Meſsier has it not.
Observed the nebula near the 2 Navis. a cluster of
bright ſtars. (Meſs. 46.)
Saw the nebula near Canis majoris, the ſtars are
not ſo numerous and bright as in the 46th
(Meſs. 41st)

Fig. 2.1 Caroline's records of her sweeps on 26 February 1783. She says of the first nebula, "Messier has it not", but it was in fact to be found in Messier's final catalogue of which the Herschels then knew nothing. The second nebula was indeed new, and the events of the evening convinced William that nebulae were present in great numbers awaiting discovery. Courtesy of the Royal Astronomical Society

found herself in the dark, perched (in a dress!) of the huge ladder, attempting to find a particular double star and then to measure the angle of the line joining its two components to see if this had changed. The whole thing was preposterous.

Fig. 2.2 The 'large' 20-ft reflector in its prime, from the engraving William published in 1794

Fortunately for Caroline, William's solo efforts were running into trouble. Standing on the observing gallery, he would drag the tube to one side, begin to look for nebulae, and slowly manhandle the tube across to the opposite side. If he succeeded in finding a nebula, he would need to use artificial light to make written notes, and he would then have to wait a little until his eyes were dark-adjusted once more and he could resume his search. By that time he could have little confidence in exactly which bit of sky it was that he had swept. It was a hopeless way to proceed.

But as usual William was quick to learn. If he was to detect the faintest of nebulae, he must dedicate himself to the actual observing, and leave someone else—Caroline—to keep the written records. And so at the end of 1783 a relieved Caroline was taken off the small 20-ft and installed at a desk in front of a window overlooking the large 20-ft. When William pulled a cord, she would open the window. He would then shout out the observations she was to record, and she would shout them back to make sure she had heard correctly.

William established the position of a nebula by its relation to a nearby star, at first along the lines of "left a bit, up a bit", later by comparison of the two sets of coordinates. Caroline therefore needed to identify the star in question. At first she used the 'bible' of observers, the British Catalogue of John Flamsteed, the first Astronomer Royal; but this was arranged by constellation, and in his sweeps William would regularly cross the boundary between one constellation and the next. She therefore compiled from the Catalogue her own reference list of stars,

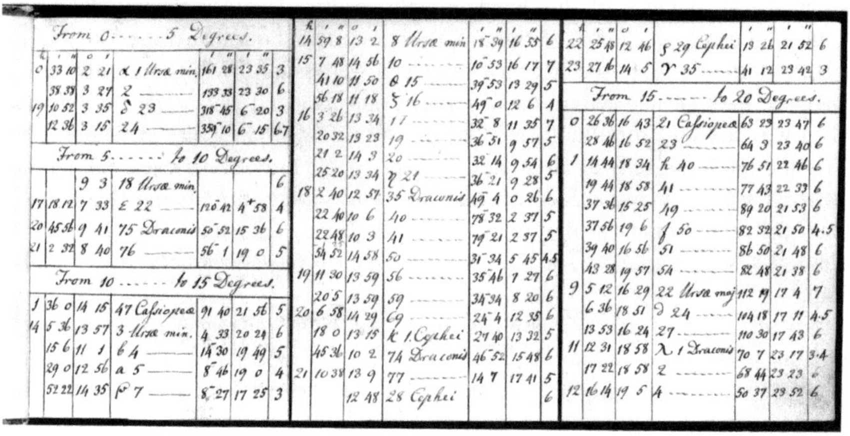

h	′ ″	° ′		′ ″	′ ″	
From 0 ...... 5 Degrees.						
0	33 10	2 21	α 1 Ursæ min.	161 28	23 35	3
	38 38	3 27	2 ———	133 33	23 30	6
19	10 52	3 35	δ 23 ———	318 45	6 20	3
	12 36	3 15	24 ———	359 10	6 15	6.7
From 5 ...... to 10 Degrees.						
		9 3	18 Ursæ min.			6
17	18 12	7 33	ε 22 ———	120 42	4 58	4
20	45 56	9 41	75 Draconis	50 52	15 36	6
21	2 32	8 40	76 ———	56 1	19 0	5
From 10 ...... to 15 Degrees.						
1	36 0	14 15	47 Cassiopeæ	91 40	21 56	5
14	5 36	13 57	3 Ursæ min.	4 33	20 24	6
	15 6	11 1	b 4 ———	14 30	19 49	5
	29 0	12 56	a 5 ———	8 46	19 0	4
	52 22	14 35	β 7 ———	8 27	17 25	3
14	59 8	13 2	8 Ursæ min	18 39	16 55	6
15	7 48	14 56	10 ———	10 53	16 17	7
	41 10	11 50	θ 15 ———	39 53	13 29	5
	56 18	11 18	ζ 16 ———	49 0	12 6	4
16	3 26	13 34	17 ———	32 8	11 35	7
	20 32	13 23	19 ———	36 51	9 57	5
	21 2	14 3	20 ———	32 14	9 54	6
	25 20	13 34	η 21 ———	36 21	9 28	5
18	2 40	12 57	35 Draconis	49 4	0 26	6
	22 40	10 6	40 ———	78 32	2 37	5
	22 48	10 3	41 ———	79 21	2 37	5
	54 52	14 58	50 ———	31 34	5 45	4.5
19	11 30	13 59	56 ———	35 46	7 27	6
	20 5	13 59	59 ———	34 34	8 20	6
20	6 58	14 29	69 ———	24 4	12 35	6
	18 0	13 15	κ 1 Cephei	27 40	13 32	5
	45 36	10 2	74 Draconis	46 52	15 48	6
21	10 38	13 9	77 ———	14 7	17 41	5
		12 48	28 Cephei			6
22	25 48	12 46	ξ 29 Cephei	13 26	21 52	6
23	27 16	14 5	γ 35 ———	41 12	23 42	3
From 15 ...... to 20 Degrees.						
0	26 36	16 43	21 Cassiopeæ	63 23	23 47	6
	28 46	16 52	23 ———	64 3	23 40	6
1	14 44	18 34	h 40 ———	76 51	22 46	6
	19 44	18 58	41 ———	77 43	22 33	6
	37 36	15 25	49 ———	89 20	21 53	6
	37 56	19 6	f 50 ———	82 32	21 50	4.5
	39 40	16 56	51 ———	86 50	21 48	6
	43 28	19 57	54 ———	82 48	21 38	6
9	5 12	16 29	22 Ursæ maj	112 19	17 4	7
	6 36	18 51	d 24 ———	104 18	17 11	4.5
	13 53	16 24	27 ———	110 30	17 43	6
11	12 31	18 58	λ 1 Draconis	70 7	23 17	3.4
	17 22	18 58	2 ———	68 44	23 23	[illegible]
12	16 14	19 5	4 ———	50 37	23 52	6

Fig. 2.3 The first page of the catalogue Caroline assembled from Flamsteed's British Catalogue, so that she could identify the stars that William encountered in his current sweep. In any given sweep, all these stars would be at similar distances from the North Pole, and so her catalogue is arranged in 'zones' of north polar distance. Courtesy of the Royal Astronomical Society

arranged by North Polar Distance, for all the objects William would come across in a sweep would be at similar angles from the North Pole (Fig. 2.3).

In the morning after a night's sweeping, Caroline would make a fair copy of the nebulae William had discovered, and in due course she would compile catalogues—two of a thousand nebulae and a third of five hundred, as it transpired—for publication by the Royal Society. She was indeed William's 'assistant', and he would refer to her as such in print, but her work needed to be meticulous in the extreme and her energies unflagging, and she played an essential role in the great endeavour.

There was a third member of the team, a humble mechanic who systematically raised and lowered the tube of the reflector. Its field of view was a mere 15 min of arc; and the rotation of the sky was so slow that William had ample time to examine a much wider zone, which he thought might be perhaps two degrees or so from its upper to its lower limit (although in this he was overoptimistic). To achieve the wider zone the mechanic raised and lowered the tube in a zig-zag motion, raising it until a bell that Alexander had devised told him that it was time to reverse the movement; this he would then do, until another bell told him to backtrack once again.

2.4 Fathoming the Milky Way

Galileo's telescope had long ago confirmed ancient speculations that the Milky Way is in fact a collection of tiny stars, but for a long time surprisingly little effort was given to the question of the three-dimensional form of the system of stars from which the appearance of this band of light results. Then, in the middle of the eighteenth century, Thomas Wright of Durham pointed out that if the Sun and the solar system are immersed in a layer of stars, we on Earth will see a milky effect when we look around within the layer, but only isolated and mainly bright stars when we look out from the layer: that is, essentially what we do in fact see. But Wright offered this only as an imagined scenario. He himself believed that our star system, the Galaxy, has in fact the shape of a vast sphere surrounding Heaven; as a result the tiny segment of the sphere which is all we can actually see through our telescopes approximates to a flattish layer—hence the appearance of the Milky Way.[8]

It is just possible that William heard talk of Wright's fantasies while he was an itinerant musician in the north of England, but it is more likely that he independently thought of the layer explanation for the Milky Way. Its stars, he said, formed a stratum, "to borrow a term from the natural historian".[9]

He was eager to establish the actual outline of this stratum, and he saw that he could achieve this if he was permitted two assumptions. The first was that his new 20-ft could penetrate to the borders of the Galaxy in all directions; although he could offer no evidence in favour of this assumption, clearly the attempted investigation was doomed from the start unless this was so.

The second assumption was more interesting. The fact that groups of stars, assembled together in clusters, were to be found in the universe strongly suggested that gravitational attraction (or a force of a similar nature) was at work among the stars, causing once widely-separated stars to move ever-closer together as they attracted each other. Now, some of the clusters William had encountered were formed of scattered stars, while others—mainly of a globular shape—were tightly packed. But the stars of a scattered cluster, *given time*, would surely continue to attract each other and so move ever-closer together, eventually evolving into a tightly-packed cluster. In other words, a scattered cluster was in the early stages of its life history, while a tightly-packed cluster was moving towards old age. This great insight of William's rang the death-knell of the essentially-unchanging clockwork universe of Newton then still prevailing, whereby God the Clockmaker had a servicing contract with His creation and intervened at times to preserve the original order; and it ushered in the modern universe in which everything, even the cosmos itself, has a life story.

If gravity worked its effects within star clusters, so would it within strata of stars, of which the Galaxy was the prime example. William's second assumption was that while in the distant future the Galaxy would inevitably fragment into isolated clusters under the continued action of gravity, this process was currently in its earliest stages: the Galaxy was in its infancy. It had started out, he proposed, by occupying a volume of space regularly stocked with stars, and these stars were

[8] See Hoskin (1970).

[9] W. Herschel, Observations tending to investigate the construction of the heavens, *Phil. Trans.*, 74 (1784), 437–451, p. 437.

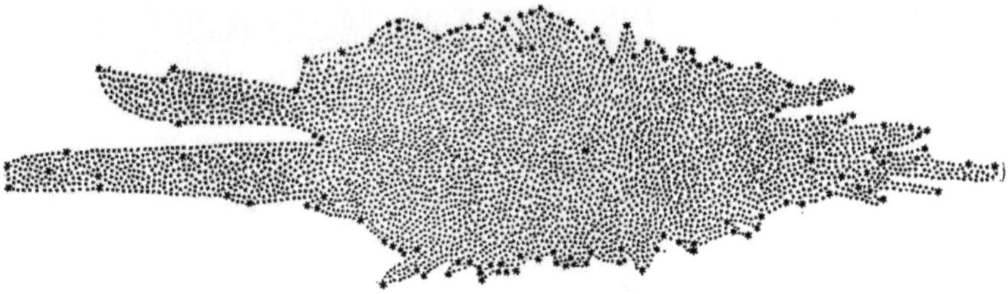

Fig. 2.4 A cross-section of the galaxy, from William's 1785 *Philosophical Transactions* paper on the construction of the heavens

still distributed with fair uniformity. This being so, the number of stars visible in any given direction was a guide to how far the Galaxy extended in that direction.

Granted these two assumptions, William was in a position to shed light on the three-dimensional shape of the Galaxy simply by counting stars, a technique that had never before been used in astronomy. He introduced it in his 1784 paper on "the construction of the heavens", a groundbreaking work that he published at a time when few astronomers were interested in what lay outside the solar system.[10] In his next paper on the construction of the heavens, sent to the Royal Society early in 1785, he set out in full detail his results for a cross-section of the Galaxy, apologising that shortage of time had prevented him from extending this work to the entire visible sky (Fig. 2.4).[11]

In the years to come he was to abandon both his assumptions. The first was disproved when he completed larger reflectors and these brought many more stars, previously invisible, into view. The second he discarded as he gained experience from his sweeps for star clusters, for he gradually realised that a high count was often due to nothing more than increased clustering in the region in question. Yet the drawing that encapsulated his results was to appear in books long after his death; for authors are like Nature and abhor a vacuum, and for decades there was nothing to take its place.

2.5 Gravity as the Agent of Change

William had swept for no more than a few weeks when he wrote his 1784 paper, but he had already encountered what we think of as clusters of galaxies. These he interpreted as older strata that had already fragmented into smaller star systems under the action of gravity. And this fate awaited our own Galaxy, which he thought might one day become a stratum of perhaps three hundred such systems.

[10] W. Herschel, Observations tending to investigate the construction of the heavens, *Phil.Trans.*, 74 (1784), 437–451, reprinted in Hoskin (2012).

[11] W. Herschel, On the construction of the Heavens, *Phil. Trans.*, 75 (1785), 213–266, reprinted in Hoskin (2012).

Fig. 2.5 William in 1785. Portrait in oils by Lemuel Francis Abbott (1760 or 1761–1802). © National Portrait Gallery, London

In a remarkable paragraph in his 1785 paper in *Philosophical Transactions*, William (Fig. 2.5) warns his readers that he would attempt to keep a balance when theorising; but that observations were of no value by themselves, and that he would speculate too much rather than too little.[12] He was as good as his word.

The fundamental question concerning the nebulae, as we have seen, was whether all of them were simply remote star clusters disguised by their distance, or whether some were clusters whereas others were nearby clouds of 'true nebulosity'. When he composed his 1784 paper on the construction of the heavens, William was satisfied that the Orion Nebula had visibly altered shape over the years, and therefore it could not be a distant and vast star cluster. Evidently some nebulae were clusters of stars, whereas others were, as we would say, gaseous—but how then could one tell the difference? William had an answer. He was convinced that some nebulae had a milky appearance whereas others were mottled, and he made the reasonable assumption that the mottled nebulae were formed of stars whereas the milky ones were truly nebulous.

But then, as is the way of such things, only days after he completed his paper his sweeps brought him first to one nebula, and then to a second, in which he could see both mottled and milky nebulosity. Both forms of nebulosity, it now seemed,

[12] W. Herschel, On the construction of the heavens, *Phil. Trans.*, 75 (1785), 213–266, p. 213.

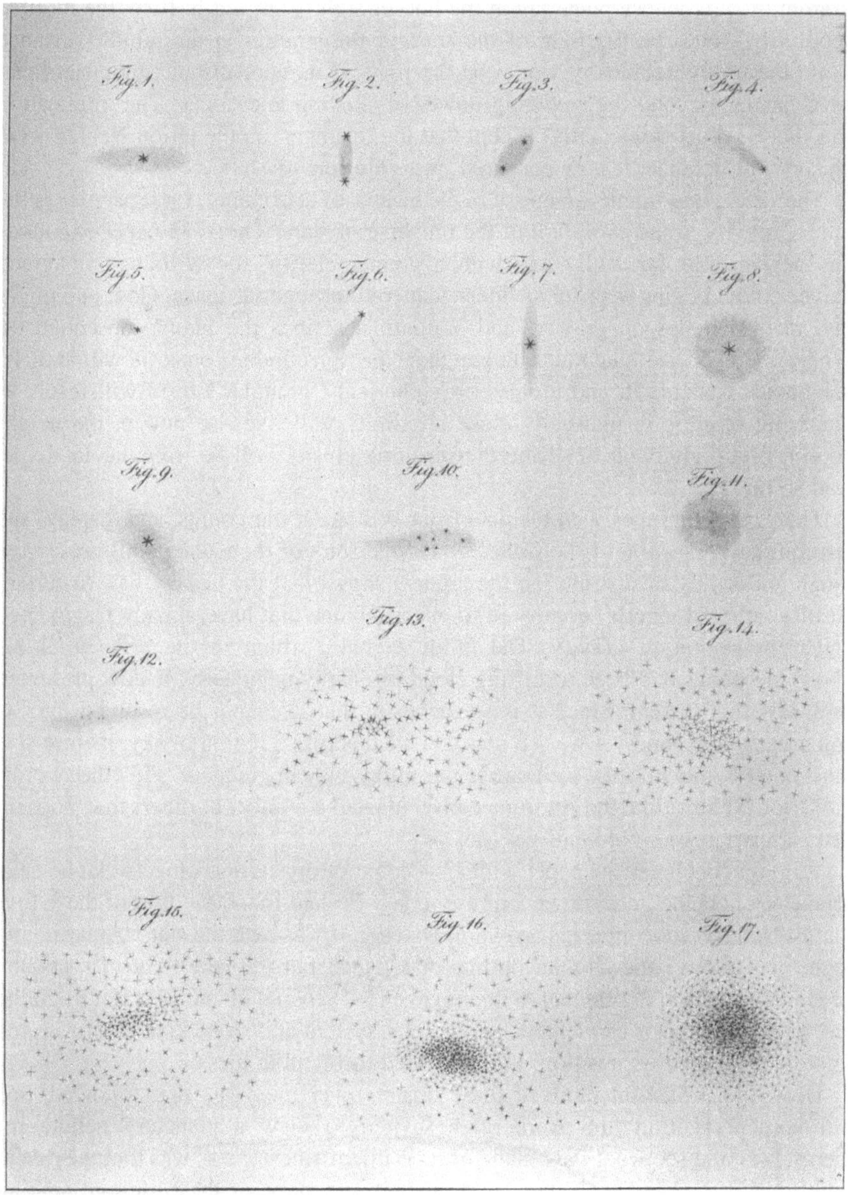

Fig. 2.6 William's conception of the evolution of star clusters under gravity, from scattered clusters to the tightly-packed globular clusters. He published this diagram in *Philosophical Transactions* in 1814 but the conception dates from the 1780s

could coexist in the same object—in which case he needed to rethink his position. The mottled nebulosity must surely reflect the presence of stars on the verge of resolution, that much was evident. But perhaps each of the two nebulae was a stratum of stars whose plane lay in the line-of-sight from Earth. If so, the mottled nebulosity would be the light of the stars of the stratum in the middle distance while the milky nebulosity would be the light of its very distant stars, stars that were far, far beyond the resolving power of existing telescopes. The implication was that he must (reluctantly) accept that the 'changes' in the Orion Nebula were illusory: all nebulae, it now appeared, were clusters of stars.

The 1784 paper marks an epoch in the history of astronomy, for it privileges the role of gravity in the evolution of the universe of stars. The 1785 paper expounds this new insight forcefully and with a greater clarity, for William's imagined universe now begins with stars widely scattered throughout space. God, one might say, then switches on gravity, and William describes the likely consequences. Where there is one star much larger than the surrounding ones, it will pull its neighbours towards it, and a cluster of spherical ('globular') form will result. A scattering of stars in unusually close proximity will likewise pull in the neighbouring stars, although this time the resulting cluster will be irregular in shape. And so on (Fig. 2.6).

The paper continues with the data from William's star counts, and displays the resulting cross-section of the Galaxy. That the stars of the Galaxy still preserve a rough uniformity of distribution (he argues) shows that the Galaxy has "a certain air of youth and vigour" compared to other nebulae that have already fragmented under the action of gravity. The Orion Nebula, which in the privacy of his observing books had been a shifting cloud of nearby nebulosity, is now presented as a vast star system. Since it is so far away that it cannot be resolved into its component stars, and yet we see it as extended across a region of sky, its true size must be vast, and it "may well outvie our milky-way in grandeur". In other words, the Orion Nebula, like the Andromeda Nebula and a number of others that William lists, are what we would call galaxies.

William concludes the paper with an account of mysterious objects that he calls 'planetary nebulae', a term we still use today. He had found the first of these (our Saturn Nebula) soon after his arrival in Datchet. It was near the star ν Aquarii, and it seemed to have the circular outline of a planet but the pale light of a nebula, hence the descriptive title that William gave it. It crossed his mind that it might be one of a new species of member of our solar system and so he kept careful watch to see if it altered its position; but it seemed to be motionless.

He found a handful more of these objects over the years that followed, and astronomers visiting him were often invited to view a planetary nebula for themselves and see what they made of it. William's best guess was that they were tightly-packed globular clusters of stars, in a late stage of their life and nearing what we would term gravitational collapse. He thought that the star that had flared up in 1572 ('Tycho's nova') might have been a planetary at the end of its life.

Chapter 3
"One of the Greatest Mechanics of his Day"

3.1 The Monster Reflector

Williams frustration at the modest size of the largest mirror at his disposal—the 18 in. of the large 20-ft—had grown steadily, as the months passed and his opinion of the nature of nebulae had shifted from one rival theory to the other. He had paid for the 20-ft out of his savings, but a reflector large enough to satisfy his ambitions would cost a king's ransom, and the only way to fund it was to hold King George to ransom.

A first step had been to allow the King to see for himself how much more detail in a nebula could be seen if one looked at it through a reflector with a larger mirror. In late August 1782 William had taken his 7-ft to the Castle and shown the King the cluster of stars that form M11, known today as the Wild Duck Cluster. A fortnight later George viewed the same object through the 10-ft, which captured nearly twice as much of the incoming light.

The next step was to refer in print, as often as possible, to the insights that a much larger mirror would bring; and here William was, as the saying goes, economical with the truth. When Watson had sent him the catalogue by Messier back in December 1781, it would surely have been natural for William to rush out that evening and view some of its nebulae through the largest of his existing instruments: the 'small' 20-ft, whose 12-in. mirrors had much greater 'light-gathering power' than the instruments that Messier used. Were these nebulae in fact "without stars" as Messier thought, or would William's reflector show them to be distant star systems?

And in his 1784 paper, this is exactly what William says he did:

> As soon as the first of these volumes came into my hands, I applied my former 20-ft reflector of 12 in. aperture to them; and saw, with the greatest pleasure, the most of the nebulae … yielded to my light and power and were resolved into stars.[1]

---

[1] W. Herschel, Observations tending to investigate the construction of the heavens, *Phil. Trans.*, 74 (1784), 437–451, p. 429.

M. Hoskin, *William and Caroline Herschel*, SpringerBriefs in Astronomy,
DOI: 10.1007/978-94-007-6875-8_3,

But William's observing books survive, and we know that this is simply untrue. As a professional musician, when Watson's present arrived he had been preoccupied with the Bath musical season then in full swing, and distracted by the prospect of royal patronage and how this might be secured. He had therefore had little time for actual observing. Surprising though it is, William neglected Messier's catalogue, with its wealth of information on nebulae that had hitherto been unknown to him, until well after he was established in his new home, many months later.

A low-key but persistent campaign by William and his allies, notably Sir Joseph Banks, culminated in the summer of 1785 in a formal application from William to Banks which Banks was to lay before the King.[2] The cost of construction of a 40-ft reflector with mirrors perhaps 4 feet in diameter was estimated at £1,395, with annual running expenses of £150:

> It appears to me highly probable that I should succeed with such an instrument, as I should not only do the whole of the optical part with my own hands, but direct every individual branch of it so as to want none but common workmen to assist me. The plan on which I should proceed would be attended with the least expense possible, since the sole end of the work would be to produce an instrument that should answer the end of inspecting the heavens, in order more fully to ascertain their construction.[3]

The King responded with a promise of £2,000, equal to the cost of construction plus running costs for the first four years.

Nothing on such a scale had been attempted before in astronomy, and George expected to go down in history as one of the great patrons of science, as indeed he was. William, desperate to put the project in motion and well aware that the size of the grant requested would be crucial in the King's decision, had shrewdly pitched his request for funding at an unrealistically modest level. Astronomers have done so ever since, confident that the granting agency, when later faced with a choice between money down the drain and the grant of further will opt for the latter course.

By this time the Herschels had given up hope of making their Datchet home waterproof. The house was still so damp that William suffered continually from the 'ague', and so in June 1785 he and Caroline moved across the river to Clay Hall in the village of Old Windsor. There they encountered a different problem—the greed of their "litigious" landlady, who demanded increases in rent whenever her tenants improved the property, even though this was at their own expense.

The 20-ft reflector could be transported from one site to another, albeit with difficulty, but the 40-ft needed a permanent home. Fortunately one of William's friends at Court was shortly coming to the end of a three-year lease on a house in Slough, and William agreed to take it over. Because Slough was on the road from Bath to London, William could easily take the coach to London for meetings of the Royal Society or whatever; and Alexander could come from Bath with his lathe to help William in telescope construction each summer when Bath was out of season.

---

[2] The story of the building of the 40-ft is described in Hoskin (2003a).

[3] W. Herschel to J. Banks, Aug. 1785, A. Aspinall (ed.), *The Later Correspondence of George III*, 1 (Cambridge, 1962), Letter 236.

The great 40-ft was only the largest of the many reflectors that would be made in the Herschel home in Slough, for after William had been a few months in post, the King had decided that he had struck too hard a bargain with his astronomer; and so he placed with him an order for five 10-ft reflectors, with the implied permission to supplement his pension in this way.[4] (Years later we find William writing to the King asking what he should do with the last of the five, which was still cluttering up his workshop—and reminding the King that he had not yet been paid for any of them.)

It is significant that William's eventual list-price for a 40-ft reflector would be 8,000 guineas, that is £8,400.[5] This included a profit margin, but the gap between this price and the £2,000 requested from the King is testimony to how far the costs of the royal 40-ft would exceed the original estimate. This would make for trouble between William and the King, trouble that is the likely reason why the internationally-famous William received no public honour until 1816, when King George's son, acting as Prince Regent (in this instance, of Hanover), made William a 'knight' of the Hanoverian Guelphic Order. Thereafter William was known as, and styled himself, "Sir William", but this was a mistake: the Hanoverian decoration was technically foreign and did not entitle a British recipient to style himself "Sir".[6]

With William and Caroline settled at Slough, construction of the 40-ft began in earnest. William would often observe all night and then spend the following day supervising the workmen. The mirror was cast in London, brought down to Windsor by boat, and then taken the short journey overland to Slough. To grind and polish a mirror of this size was a new challenge, and William had teams of a dozen men each; these wore overalls with numbers on the back, so that he could give individual workmen the appropriate orders. The workmen would unhelpfully sing and gossip, but sadly we are not told the topics of their conversations. On a later occasion a horse was recruited, but it was not a success.

The metal alloy needed for a telescopic mirror was expensive, and William had therefore opted for mirrors as thin as possible, two half-ton mirrors at £250 each in his original application. Unfortunately in casting the first mirror the mould had shrunk. In consequence the mirror was curved at the back (as well as the front), and so it did not keep its shape but distorted under its own weight as it was tilted in the tube. There would be nothing for it but to cast the second mirror thicker, and therefore even more expensive.

And so the costs of the enterprise mounted, and by the summer of 1787 it had become clear that the grant of £2,000 would be nowhere near sufficient. William was caught up in the myriad problems of construction, and he saw money as the least of his difficulties. Distracted, his usual discretion and tact deserted him, and on 18 July he wrote to a Court official, asking whether he should make make a second formal application through Banks, or whether he should simply send each of the supplementary bills as they fell due.

---

[4] Hoskin (2008a).

[5] W. Herschel to W. Shairp, 9 March 1794, RAS Herschel Archive W.1/1, 202–206.

[6] On William's claim to the appellation "Sir" see Hanham and Hoskin (2013).

On 17 August 1787 the King descended on Slough in person to see how the great reflector was progressing. He was accompanied by a vast retinue that included the Archbishop of Canterbury. The tube of the reflector was lying on the ground, and it measured well over 4 ft in diameter. The King made to walk through it—a common entertainment for William's visitors around this time—and when the Archbishop hesitated to follow, he encouraged him with "Come, my Lord Bishop, I will show you the way to heaven!"[7]

George III left William with the impression that he could count on the extra money he needed, and a few days later William went ahead with the formal application, which included £500 to cover the cost of a one-ton mirror. This proved to be a serious misjudgement. The frugal King, always resentful if he thought he was being taken advantage of, was nearing the time when the first clear signs of his eventual 'madness' would show themselves, and perhaps that was at the root of the problem. At all events, a stormy encounter ensued between the King and his astronomer, who was granted the money he had requested—but every penny was to be accounted for, even to the beer and fire needed by the workmen who would manoeuvre the reflector at night; and under no circumstances would any further funds be provided. Caroline was embarrassed and mortified; and when William's son John came across the few surviving documents many years later, he left instructions that they were to be sealed and not opened until half a century after his own death.

A few months later, the magnum opus (as William Watson always termed it) was at last nearing completion. It happened that around that time, the ring of Saturn was edge-on to the terrestrial observer, something that occurs every fifteen years. This rare configuration would give astronomers the chance to look for undiscovered moons that were normally lost in the glare of the ring, and in November 1786 Jerôme Lalande had written to William urging him to seize the opportunity.

The following August and September, William used the 20-ft to track the complex ballet of Saturn's known moons on no fewer than seventeen nights, observing on 19 August for three hours on end. He suspected he had glimpsed a sixth moon, but it was difficult for him to confirm this in the weeks that followed because of the current difficulty of distinguishing between moons and the innumerable background stars then present.

He tried again in the summer of the following year, 1788, observing on one occasion for no less than five hours. He was hampered in his search because around this time the area of sky he was currently sweeping for nebulae was near the zenith. While this was happening, the tube of the 20-ft was almost vertical, and so the mirror was horizontal. This allowed dew to collect on the mirror, which consequently tarnished more rapidly than usual.

A year later, and conditions for the study of Saturn were ideal. The ring was almost exactly edge-on, the planet accompanied by its moons was moving rapidly against the background stars, and Lalande had sent him new tables of the motions of the five known moons. For six nights William observed the planet with the

[7] Mrs J. Herschel (1879), 309.

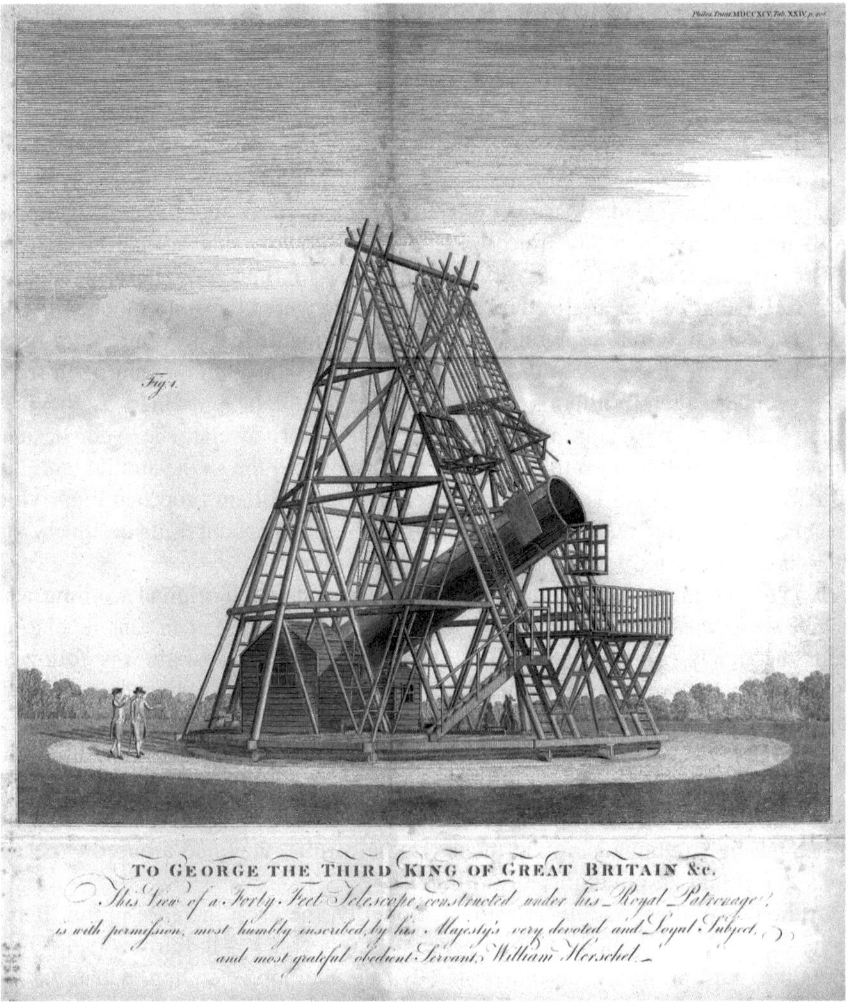

Fig. 3.1 The 40-ft reflector, as illustrated by William in the long account in *Philosophical Transactions* in 1795. Note the dedication to King George, whose glorification as a patron of science was one of the purposes of the telescope

20-ft, and then, on 28 August 1789, he was at long last able to launch the great 40-ft reflector (Fig. 3.1) on its scientific career. He directed it towards Saturn. One of the moons was known to be invisible at the time, but William could see the other four, and aligned with them what looked to be a sixth. "What makes me take it immediately for a satellite", he noted, "is its exactly ranging with the other four and the ring".[8]

[8] RAS Herschel Archive W.3/1.8.

William's hurried letter to Banks announcing the discovery was received with delight mingled with relief, for Banks had played a major part in securing both grants from the King and his own good-standing was at stake. William asked him to add to a paper he already had in press the words: "P.S. Saturn has six satellites. 40 ft reflector".[9]

But it then occurred to William that he might not be the only observer whom Lalande had persuaded to search for a sixth moon; if so, there was danger that word might arrive any day from the Continent showing that he had been forestalled. He therefore wrote a very private letter to Banks, declaring—"to my surprize"—that he now realized he had in fact discovered the sixth moon with the 20-ft back in August 1787, although other commitments had prevented him from following up the discovery at the time. If no claim to the sixth moon arrived from the Continent, then William's 1789 discovery would be presented as the first triumph of the newly-commissioned 40-ft; but if a rival claim arrived, he and Banks would jointly affirm that William had discovered the sixth satellite with the 20-ft two years before. To cover both eventualities, William proposed modifying the announcement to read: "Saturn has six satellites. An acount of its discovery, its Revolution and orbit will be given in the next volume…"[10]

No rival claim from the Continent did appear, and so the original wording was retained and the sixth satellite of Saturn was presented as an instant revelation from the newly-commissioned 40-ft. And, remarkably, the sixth was followed immediately by a seventh. Whether the King was duly impressed we do not know, for George tends to fade out of the story around this time; but Banks certainly was, and he and indeed all European men of science expected such revelations to continue on a monthly if not weekly basis. True, William—and Banks—had desperately needed something to show for the huge investment of money and effort; but by so dramatising the monster's first fruits, they had aroused expectations that it had no possibility of fulfilling.

Although William had never said so explicitly, the great increase in the 'light-gathering power' available to him when the reflector's 4-ft mirrors came into operation was primarily designed to advance the long-standing debate as to whether all nebulae were, or were not, star clusters disguised by distance. But the debate was ended once and for all, as far as William was concerned, within months of the completion of the 40-ft; and not by the 40-ft but by the 20-ft. On 13 November 1790 William was sweeping as usual with Caroline at a nearby window, when there came into view a ghostly object that was surely a star surrounded by a spherical shell of nebulosity (Fig. 3.2).

The object, known to us as NGC 1514, was in fact another planetary nebula; but this example is so close to Earth that William was able to see the central star that in fact is always present in a planetary nebula. Accordingly, William classified the object as a nebulous star. Marvelling at what he saw, he accepted the evidence

[9] W. Herschel to J. Banks, 29 Aug. 1789, Lubbock (1933), 163–164.

[10] W. Herschel to J. Banks, 4 Sept. 1789, Lubbock (1933), 164–165.

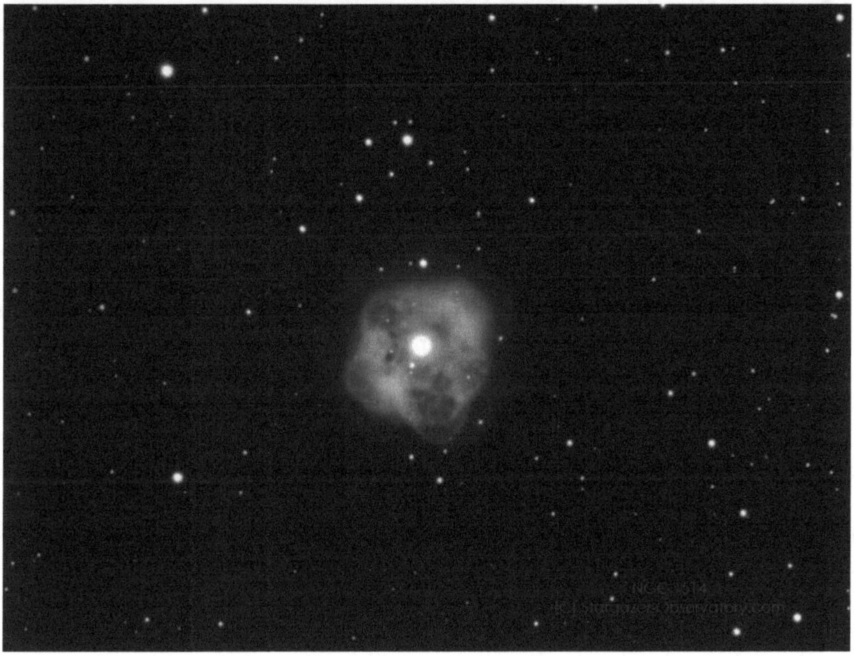

Fig. 3.2 The planetary nebula now known as NGC 1514, which William encountered on 13 November 1790. He saw it as a central star condensing under gravity from the surrounding nebulosity, and this convinced him that 'true nebulosity' existed after all. Image courtesy of Dietmar Hagar, FRAS © www.stargazer-observatory.com

before his very eyes, even though it contradicted the theory he had argued in his last two 'construction' papers. He had to accept that the spherical glow around the star could not be a distant globular cluster that by pure chance lay in exactly the same direction as the star: it must be 'true nebulosity' out of which the star was condensing—growing.[11] A star could then be said to have a "biography".[12]

He had therefore been wrong to argue that all nebulae were star clusters, and wrong to abandon as illusory the changes that he had (supposedly) observed in the Orion Nebula. In later papers on the construction of the heavens he would spell out the implications of this new insight, but the immediate effect—although he would not admit this in public—was to deprive the 40-ft of its primary purpose.

The monster's secondary purpose was to glorify its royal patron. And so its cumbersome structure—which required not one but two workmen to manoeuvre—had to be maintained in some semblance of working order for the rest of William's life, to satisfy the endless stream of aristocratic visitors sent to Slough from Windsor Castle.

---

[11] W. Herschel, On nebulous stars, properly so called, *Phil. Trans.*, 81 (1791), 71–88, reprinted in Hoskin (2012).

[12] W. Herschel, On the periodical star α Herculis, *Phil. Trans.*, 86 (1796), 452–482, p. 457.

3.2 Attraction by an Earthly Star

Of the five 10-ft reflectors that King George had ordered, one had been destined for Göttingen University, a favourite 'good cause' of his in his role as Elector of Hanover. With no thought for the disruption this would cause to the work on the 40-ft, George had instructed William to go to Germany and deliver the instrument in person. So off William went in July 1786, accompanied by Alexander, while Alexander's new wife spent the time at Slough, to infuriate Caroline by her gossipping.

With William absent, Caroline was free to pursue her own observing. The first little sweeper that William had made her was most inconvenient, for as the tube was rotated around the vertical spindle, Caroline the observer had to accompany it, in a tiresome cyclic dance. In the summer of 1783, therefore, William had devised for her a much more convenient instrument. It was a Newtonian reflector, and the tube was designed to pivot about its upper end in a vertical plane, first from the zenith to the horizon and then back again, in response to rotations of the handle by the observer. Being a Newtonian, its eyepiece was also at the upper end. This eyepiece would of course rotate with the tube into which it was fixed, but otherwise it was completely still. As a result, when looking through it Caroline could sweep a vertical strip of sky without so much as moving her head: she turned the handle, and saw the sky slowly change before her eye.

Happily, it often turned out that by the time she had examined one vertical strip of sky to her satisfaction, the rotation of the heavens had already brought the adjacent vertical strip into view; if so she could sweep simply by rotating the telescope up and down in regular fashion. The contraption was ingenious, and Caroline would one day take it back to Hanover in her retirement, as a keepsake.

There was no longer any point in Caroline's searching for nebulae, for she could not compete with the 18-inch mirrors of the 20-ft. As for double stars, William's catalogues contained all the stars of this type that one could wish for. Caroline therefore committed herself to searching for comets, newly-arriving visitors from outer space for which the wide field of view of her new sweeper was ideal. And in William's absence, on 1 August 1786, she found her first comet.

She wrote immediately to Dr Charles Blagden, secretary of the Royal Society, for if professional astronomers were to make an accurate determination of the comet's orbit they must be alerted to its arrival as quickly as possible. By a happy chance her letter reached Blagden just before the annual meeting of the Visitors of the Greenwich Observatory, when the principal astronomers of the land assembled together, and so Blagden was able to give them the news in person.

Not only that, but five days later Banks and Blagden—the president and secretary of the Royal Society—along with Lord Palmerston journeyed to Slough, for the express purpose of seeing Caroline's comet through Caroline's telescope. The scullery in Hanover must have seemed a long way away.

Fig. 3.3 Miniature on ivory of William's wife Mary, painted in March 1805 by John Keenan, who was soon to become court painter to Queen Charlotte. Herschel family archives

William returned to learn of his sister's success, and he was summoned to Windsor Castle so that the Royal Family and their friends might see "the first lady's comet".[13] It would not be the last.

Caroline was enjoying a happy and fulfilled life. Her brother's sweeps for nebulae had already resulted in their first catalogue of one thousand such objects, and a second such catalogue would follow shortly. She herself was playing an essential role in this great enterprise, for how else could William record his observations, and who else would write them up for publication with the impeccable accuracy that was essential? In whatever time was left to her she ran his household and entertained his guests—and even did some observing on her own account.

But there was a cloud on the horizon. The landlord of their Slough home was a widow, Elizabeth Baldwin, who lived nearby. She came from a wealthy family, and she had married into a still wealthier family: the Baldwins, who had made their money from the coaches that carried the aristocracy between London and Bath, and from the coaching inns they built along the route. Elizabeth's daughter Mary (Fig. 3.3) had married John Pitt, himself a wealthy 'gentleman' who did not need to work but who lived off the rents from his properties.

---

[13] Charlotte Barrett (ed.), *Diaries & Letters of Madame D'Arblay* (London, 1905), entry for Aug. 1786.

John and Mary had their home in the next village, Upton, but their property (which was owned by Elizabeth) shared a boundary with that of the Herschels. The two families became friends, and when John died in the autumn of 1786, the friendship between William and the Widow Pitt became altogether too close for Caroline's liking. Their courtship led to engagement to marry. But William was not willing to abandon Caroline who was essential to the observing program at Slough, while Mary expected her future husband to spend the bulk of his time with his wife at Upton, rather than with his sister at Slough. Mary therefore broke off their engagement, no doubt to Caroline's huge relief. But after a while negotiations were reopened, and in the summer of 1787 a curious deal was done whereby William and Mary were to maintain both their current establishments, and employ a footman to go between them. Nor was this the end of the bizarre episode, for early the following year William commissioned Watson to take soundings from members of the astronomical community to see whether his marriage would meet with their approval. Fortunately the consensus was that William's astronomy might actually benefit if he was a little less intense about it.

And so, on 8 May 1788, William and Mary were married in Upton parish church. Sir Joseph Banks was best man, Alexander and Caroline were witnesses, and Mary, because her father was dead, was given away by her brother Thomas. The reception and the ensuing honeymoon took place in the newly-weds' Slough home, but it was from Upton that the customary cards were sent to friends a few weeks after the wedding. This was the beginning of an attempt to maintain the twin households as agreed, and during this time guests invited to dinner sometimes found they had turned up at the wrong house. But by September 1791 Mary, who was then pregnant, had moved permanently to Slough and was renting out her Upton property.

William's marriage marked the end of the years when Caroline had contentedly presided over his household. She moved out of the main house at Slough and into the adjoining cottage, where she had an excellent rooftop site on which to position her comet-sweeper. She even had a little money to call her own. William, who was marrying a woman rich beyond the dreams of avarice, had offered his sister a regular payment for her work as his assistant; but Caroline was tired of brotherly handouts, and in a surprising show of spirit insisted that William ask the King (or Queen) for her to receive in future a 'pension' of her own. And so Caroline, with £50 a year promised from the royal purse, became the first female professional astronomer in history.

A Herschel would always deal with setbacks by destroying any written evidence and getting on with his or her life. Caroline's world fell apart when her adored brother chose to give another woman priority over her, and her bitterness left its mark in her diary. But Mary proved to be a charming wife and sister-in-law, who brought William great contentment, and when Caroline later came to write her autobiography, she regretted the anger that she found expressed in her diary and simply burnt it.

William now had a purpose in life outside astronomy, and the stars had to compete with Mary for his attention at night time. After 1790, when the problem

Fig. 3.4 The 25-ft reflector that William sold to the King of Spain in 1802, in practical terms the most successful of his great reflectors. This water-colour was part of the instructions for assembly. The mounting was destroyed by Napoleonic troops in 1808, but a full-scale replica has recently been completed. Courtesy of the Observatorio Astronómico Nacional, Madrid

of the nebulae was resolved to his satisfaction, he began to lose interest in sweeping. He and Caroline already had over two thousand nebulae in the bag, so why bother with more? Thereafter they made such slow progress with the remaining sweeps that it was not until 1802 that they had enough nebulae for a third catalogue, and this time of five hundred only. Caroline's immaculate charts of the regions swept survive for the sky more than 30° from the Pole, and they show that only about one-twentieth of this area still remained unexamined. The zenith at Slough is at 38½° North Polar Distance, and to examine regions closer to the Pole it was of course necessary to rotate the 20-ft reflector so that it faced north (and then to check that its alignment was exact), so there was a disincentive to tackle the polar region as long as there was work to do to the south of the zenith. In the event William swept only about half of the sky with less than 30° NPD, none of it within 6° of the Pole.

Nevertheless, in 1802 the ambitious project to sweep the entire sky visible from Slough was within sight of conclusion, and a few weeks of determined effort would have been sufficient; but William simply left the remaining areas of sky unswept.

William's trade in reflectors had long since become a minor industry. At first it had brought in much-needed additional income, but it had also allowed him to

Fig. 3.5 Portrait in oils of John Herschel aged seven painted in 1799 by Robert Muller (1773–c. 1800). Windsor Castle, a couple of miles from the Herschel home, is shown in the background. Herschel family archives

make experiments in grinding and polishing large mirrors. In the years to come William was to delight in his reputation as "one of the greatest mechanics of his day",[14] and the crowned heads of Europe were reduced to begging him to accept their commissions, knowing that if he declined there was no-one else to whom they could turn. And so we find William, long after he had become rich by marriage, making a 25-ft reflector for the King of Spain (Fig. 3.4), and a 20-ft for Catherine the Great of Russia.

He also provided Britain with an early-warning system against invasion from France, a 7-ft reflector that was installed in 1799 in Walmer Castle overlooking the Straights of Dover. It arrived in kit form with a dozen pages of instructions for assembly, and the Prime Minister, William Pitt, himself made an attempt to follow them.

> Mr Pitt perused your instructions with satisfaction [wrote Charles Burney to William]; & when he came to the six short rules laid down—to the first 2 he said: "we are right thus far—but the putting on the small speculum [mirror] has puzzled us"—the rest he seemed to think not difficult. And said that he had left a person at work, who seemed to understand the business.[15]

After his marriage, William began to take leisurely tours, often with Mary and their son John (Fig. 3.5) who was born in 1792, sometimes with other friends, but

[14] C. A. Lubbock, author's typescript of *The Herschel Chronicle*, chap. 11, p. 28, William Herschel Museum, Bath, citing the journals of Mrs Charlotte Papendiek.

[15] C. Burney to W. Herschel, 3 Sept. 1799, RAS Herschel Archive W.1/13.B.178.

never with Caroline. She was paid by the King to be his assistant astronomer, but he regarded her simply as his assistant, and when William was on tour she would be left behind to 'mind the shop'. Happily, with the sweeps in the doldrums, she was only rarely needed at night as amanuensis, and this allowed her to make a career for herself, as a comet-hunter.

Chapter 4
The Peerless Assistant

4.1 "Priestess of the New Heavens"

The comet-sweeper that William had made Caroline, with a tube that was little over two feet in length, was one that his sister could use while she was sitting comfortably alongside, observing through the eyepiece located near the top of the tube. But William was never one to leave well alone, and in 1790 he made her a bigger and (in his eyes) better one, of five feet focal length. Since the new tube was taller than Caroline, she now had to stand on a stool in order to peer through the eyepiece at the top of the tube. This she found tiring, and sometimes she would opt for her old friend the 'small' sweeper.

Caroline had already discovered one comet, when William was away in Gottingen, and in the decade 1788–97 she found no fewer than seven more. She discovered the first in December 1788, when observers were on the lookout for a possible return of the great comet of 1661, because the 1661 comet was suspected to have been itself a return of the comet of 1532. Caroline's searches were interrupted by nights when she had to act as amanuensis to William, but on 21 December she found a comet near the star β Lyrae. It was not in fact the comet of 1661 but one previously unknown; Caroline's comet returned in 1939 and is expected again in 2092.

Maskelyne was hugely impressed by the discovery. "As it came up from the south", he wrote to William, "it seems that Miss Herschel lost no time in finding it. I mean that it could not have been seen much sooner even in her excellent glass." It was, he declared, a "Cometic Ghost, conjured up by your sister, with her powerful glass instead of a wand".[1]

It took Caroline just over a year to find her third comet, but the fourth followed only a few weeks after, on 17 April 1790. Back in 1783, and again in 1787, William had seen bright spots on the Moon, and these he had interpreted as active volcanoes. From time to time thereafter, Caroline examined the Moon to see if

---

[1] N. Maskelyne to W. Herschel, 10 Jan. 1789, RAS Herschel Archive W.1/13.M.37.

DOI: 10.1007/978-94-007-6875-8_4,

there was any further evidence of volcanic activity, and she chanced upon her fourth comet while thus engaged.

As always, Caroline saw her work as done when she had alerted professionals to the visitant. When she discovered a comet she never took the slightest interest either in its orbit or in its physical nature: like a gun dog, her task was to bring the trophy home. But on this occasion the letter was slow to reach Maskelyne, and this and subsequent inclement weather cost him six nights of viewing the comet. Nor could Caroline's original observations supply the data he needed, for her way of specifying the comet's location left much to be desired. He begged her next time to send word as quickly as express post would allow; he would be more than pleased to meet the cost.

Caroline found further comets in 1791, 1793 (although it transpired that this time she had been anticipated by Messier), and 1795. The 1795 comet was to prove especially interesting, for a quarter of a century later the German astronomer Johann Franz Encke would show that what Caroline had witnessed was a return of the comet previously seen by a French observer nine years before, and that this object (now known as 'Encke's Comet') in fact reappeared with the unprecedented frequency of every three years.

These discoveries brought Caroline international fame (Fig. 4.1), and Continental male astronomers vied with each other to devise for their improbable lady colleague ever more flowery compliments. Thus Professor Karl Felix Seyffer of Göttingen wrote to ask: "Permit me, most revered Lady, to bring to your remembrance a man who has held you in the highest esteem ever since he had the good fortune to enter the Temple of Urania, at Slough, and to pay his respects to its priestess…."[2]

Caroline discovered her eighth comet with the naked eye, on 14 August 1797. It happened that William was away, and so she had to use her own initiative to get word urgently to Maskelyne. She opted for direct action. She allowed herself a short nap, and then saddled a horse and rode all the way from Slough to London and then on from London to Greenwich, where she arrived on the doorstep of an astonished Maskelyne, exhausted but with the precious observations.

She stayed with him a night or two, and then set off for home. Maskelyne urged her to stop off in London and give the news of the comet to Sir Joseph Banks, but Caroline was shy, even with an old friend like Banks, and she preferred to ride straight to Slough. She then penned Banks a note of apology:

> I thought a woman who knows so little of the world ought not to aim at such an honour, but go home, where she ought to be, as soon as possible.[3]

Caroline laid claim to eight comets, but she may unknowingly have discovered a ninth. Back in the summer of 1783, when she was proudly discovering nebulae unknown to Messier, she twice records seeing in the tiny constellation of Equuleus

---

[2] Mrs J. Herschel (1879), 92.

[3] C. Herschel to J. Banks, 17 Aug. 1797, Mrs J. Herschel (1879), 94–95.

Fig. 4.1 Etching dated February 1790 showing "The Female Philosopher smelling out the Comet" (which is in fact a meteor!). Caroline had found a comet the previous month. From a private collection

"a rich spot". These are the only two occasions when she uses this term, and there are no plausible nebulae in the locations in question. It seems likely that she had seen a comet that had escaped all other observers (Messier was still recovering from a fall) and which has been unknown to astronomers until now.[4]

4.2 The British Catalogue Revised

By the late 1790s, with Mary running the Herschel home and with sweeps becoming a rarity, Caroline had time on her hands. William therefore asked her to undertake a major piece of deskwork that called for someone with Caroline's dedication and impeccable accuracy: a revision of John Flamsteed's British Catalogue of stars.

In his early days as a professional astronomer, William had occasionally come across a star whose magnitude was not as listed by Flamsteed, or which was simply not present in the catalogue. Other stars included in the catalogue were no longer to be found. Such was the reputation of Flamsteed that at first it did not occur to William that the catalogue might be in error, and so he invented far-fetched explanations for changes that had apparently occurred between Flamsteed's time and his: thus a star might have moved directly towards the solar system and increased in apparent brightness as it did so—although it seemed odd that the star should move exactly in our direction. Eventually it dawned on him that Flamsteed's catalogue, magnificent though it was, was not free from error.

The problem was that the catalogue itself formed vol. iii of Flamsteed's multi-volume work, while the observations on which the catalogue entries were based were in vol. ii; and there was no way of going back from the entry in vol. iii to the corresponding observation in vol. ii, to see whether an error of transcription had crept in. In the autumn of 1795, therefore, William "invited" Caroline to generate an index to correlate the catalogue entries with the observations. The task took her twenty months of methodical, painstaking work, at the end of which Caroline had uncovered hundreds of errors in the catalogue. She had also identified five hundred stars that Flamsteed had omitted by oversight: the British Catalogue should have contained, not three thousand stars, but three thousand five hundred.

This revision of Flamsteed, for which Caroline's talents made her uniquely qualified, restored the reliability of the British Catalogue, and at Maskelyne's urging the Royal Society paid her the exceptional compliment of publishing it as a bound volume at their own expense.[5] It runs to seventy-six folio pages of numbers, and if there is a single slip, it has yet to be identified (Fig. 4.2).

---

[4] Hoskin (2005a), 400–402.

[5] C. L. Herschel, *Catalogue of Stars, Taken from Mr. Flamsteed's Observations...* (London, 1798).

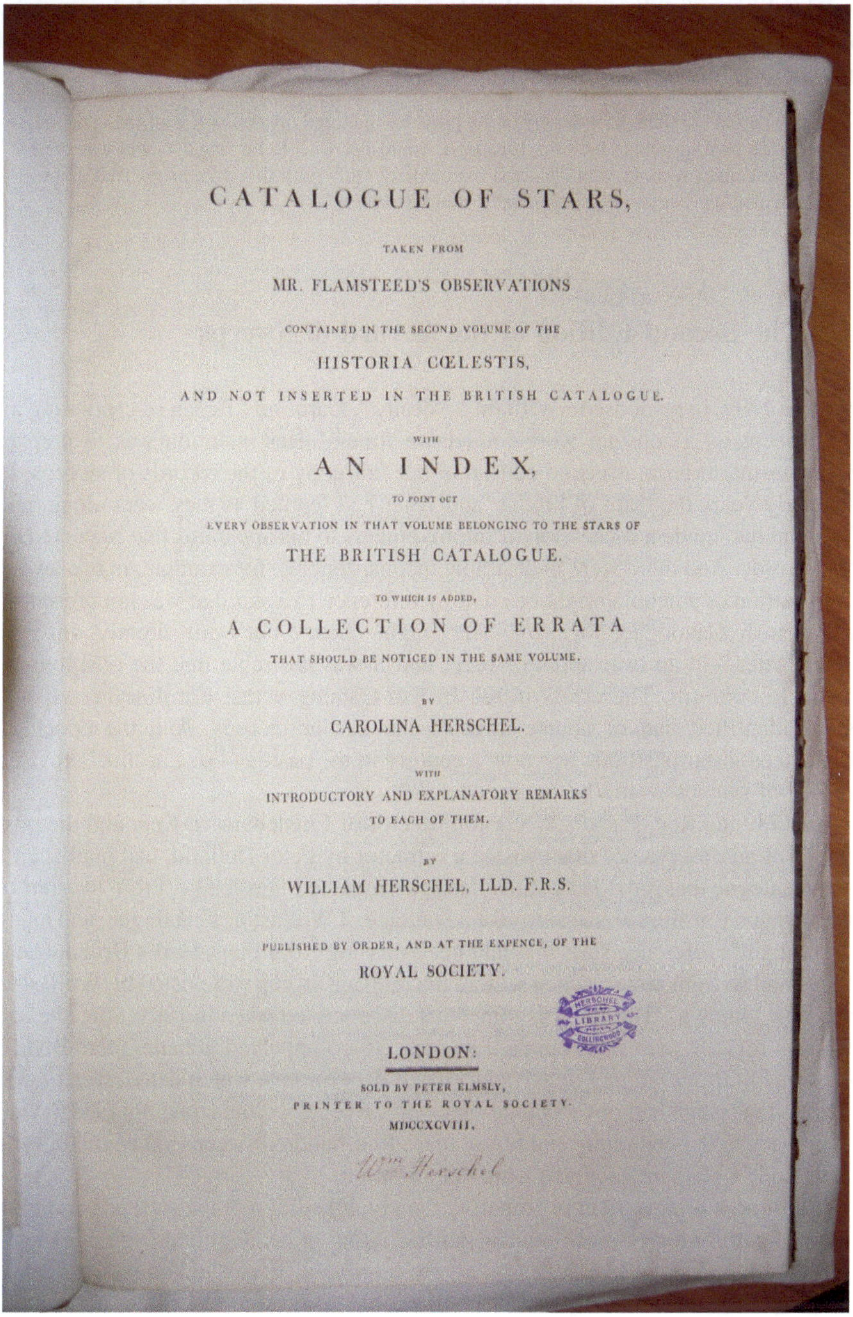

CATALOGUE OF STARS,

TAKEN FROM

MR. FLAMSTEED'S OBSERVATIONS

CONTAINED IN THE SECOND VOLUME OF THE

HISTORIA CŒLESTIS,

AND NOT INSERTED IN THE BRITISH CATALOGUE.

WITH

AN INDEX,

TO POINT OUT

EVERY OBSERVATION IN THAT VOLUME BELONGING TO THE STARS OF

THE BRITISH CATALOGUE.

TO WHICH IS ADDED,

A COLLECTION OF ERRATA

THAT SHOULD BE NOTICED IN THE SAME VOLUME.

BY

CAROLINA HERSCHEL.

WITH

INTRODUCTORY AND EXPLANATORY REMARKS

TO EACH OF THEM.

BY

WILLIAM HERSCHEL, LLD. F.R.S.

PUBLISHED BY ORDER, AND AT THE EXPENCE, OF THE

ROYAL SOCIETY.

LONDON:

SOLD BY PETER ELMSLY,

PRINTER TO THE ROYAL SOCIETY.

MDCCXCVIII.

Fig. 4.2 William's personal copy of Caroline's *Catalogue of Stars, taken from Mr. Flamsteed's Observations*… (London 1798). Courtesy of the Royal Astronomical Society, photograph by Jennifer Higham

Caroline was understandably delighted with this recognition. Maskelyne was by now a close friend who looked on her as a fellow-astronomer to be cherished, and in thanking him she did not hesitate to pull his leg:

> But your having thought it worthy of the press has flattered my vanity not a little. You see, Sir, I do own myself to be vain, because I would not wish to be singular; and was there ever a woman without vanity?—or a man either? Only with this difference. That among gentlemen the commodity is generally stiled ambition.[6]

4.3 The Second Edition of the Record of Sweeps

In June 1799, in response to William's "desire", Caroline "Began re-caculating all the sweeps as a constant work for leisure time".[7] Her intention was to prepare what we might term a second edition of her fair copy of the records of sweeps. In the early years the team of brother and sister had learned as they went along, and William had made a succession of improvements to his apparatus that impacted on the records. And there were undesirable inconsistencies: for example, in one sweep the position of a nebula might be cited by reference to a star that was not present in the British Catalogue; the same star might occur—but with slightly different coordinates—in an overlapping sweep, and it was desirable that the positions be made to conform. The errors in the British Catalogue that Caroline herself had since identified had of course infected her earlier records. And the epoch of Flamsteed's star positions was now a century in the past, and so Caroline's revised record of sweeps would be for epoch 1800.

In 1789 the Rev. Francis Wollaston, Rector of Chislehurst in Kent and the possessor of an observatory that boasted a refractor by Peter Dollond, had published a star catalogue that provided a valuable supplement to Flamsteed's. From the start of her revision Caroline was able to take advantage of Wollaston's catalogue, and many of William's reference stars that had been missing from Flamsteed's British Catalogue and so "unknown" when seen in the relevant sweep were listed by Wollaston and now "known". But she was somewhat disconcerted when in 1801, after she had already revised over four hundred sweeps, there appeared Johann Elert Bode's *Allgemeine Beschreibung und Nachweisung der Gestime*, which listed over 17,000 stars, all of which had now become "known". Even Caroline could not face having to go back to the beginning and revise these four hundred sweeps yet again, but she took Bode's stars into account from that point on.

Caroline's achievement in preparing the second edition of the sweeps is nothing short of extraordinary. Each double star had either to be identified from one of the earlier catalogues, or else noted as a new discovery. Each "unknown" reference star had to be checked against stars listed by Wollaston and Bode, and where

[6] C. Herschel to N. Maskelyne, Sept. 1798, Lubbock (1933), 257.

[7] Mrs J. Herschel (1879), 100.

necessary its position calculated for the epoch 1800. Each nebula had to be checked to see if it had already been observed in an earlier, overlapping sweep, and if not assigned the next serial number. And if she had to put the work to one side for a while when other matters intervened, it would be a major challenge then to refresh her memory of how the calculations were to be carried out.

Her new edition of the data from 1112 sweeps was concluded at last on 16 March 1804: "Finished re-caculating sweeps.... Above 8,760 observations have been brought to [the year] 1800".[8] The records of two decades of sweeping were now in the immaculate order that is the hallmark of Caroline's deskwork.

4.4 A Family Spat

In the mid-1790s, Caroline was comfortably housed in the cottage adjacent to the main Herschel residence. She had ample time and an ideal setting in which to pursue the searches for comets that were bringing her fame, she was immediately on hand if her brother (Fig. 4.3) decided to make one of his increasingly rare sweeps, the books she needed for her desk work were in the library, and she had her little nephew John on whom to lavish her maternal instincts. But then something went wrong. What happened we can only guess, for she later destroyed her diary entries for this period, but perhaps either William or Mary—very uncharacteristically—made reference to her status as their lodger. Caroline now had a salary of her own, and the decades of swallowing her pride during family rows were past. Whatever the reason, in October 1797 she suddenly asked William's workman Sprat if she might take lodgings in his home.

Sprat was no doubt delighted, and Mrs Sprat equally so when asked—for a consideration—to clean and cook for their lodger. The new arrangements worked reasonably well, for if William decided to sweep he would need both Caroline and Sprat, Caroline to take notes and Sprat to manoeuvre the 20-ft; and Sprat could later escort Caroline home. If on the other hand William and Mary decided to go off to Bath, or to their favourite holiday resort of Dawlish on the south Devon coast, Caroline would go each day to the observatory and do her paperwork, and if the night was fine she would search for comets and afterwards Sprat would come and see her home.

After two years with Sprat, at Michaelmas 1799, Caroline took rooms with a "sober, industrious" tailor who lived in the High Street. In later life the tailor's son would recall the times when Caroline knocked on the wall between their bedrooms, because the skies had cleared and William would need her attendance. The boy would dress and come downstairs to find Caroline waiting. "Will you take me to my Broder?",[9] she would say, and he would escort her to the observatory.

8 Mrs J. Herschel (1879), 109.

9 Lubbock (1933), 296.

Fig. 4.3 Pastel portrait of William in 1794 by John Russell (1746–1806), showing him with Uranus ("The Georgian Planet") and two satellites. Herschel family archives

Sober and industrious the tailor might have been, but whatever Caroline paid him it was not enough to meet his financial needs, and before the end of 1799 there was a disturbing episode in which "The bailiffs took possession of my landlord's goods, and I found my property was not safe in my new habitation".[10]

Back in February of that year William and Mary had departed Slough for Bath, where they had rented a house on Sion Hill, some way out from the centre. That month the Bath *Chronicle* announced "We hear Dr Herschel is about to make this place his residence for eight to ten months of the year. His apparatus is already fixing at his intended abode on Sion Hill".[11] He rejoined the Bath Philosophical Society, which had lapsed but recently been revived, and subscribed to a music library. Between February and October he spent more than half his time in Bath.

On a journey through Yorkshire back in 1793, William had called at the home of John Michell, who had then recently died, and arranged to purchase Michell's tools along with parts of Michell's great reflector. The mirror had been a remarkable 29 inches in diameter, but it had been little used and was now cracked.

[10] Mrs J. Herschel (1879), 104.

[11] On William's attempt to establish a second home in Bath see Hoskin (2012a).

William's more modest 20-ft with its 18-inch mirrors was ideal for sweeping, but the sweeps were coming to an end. True, the 20-ft could be used for other work, but William always needed a workman on hand to point it in the required direction. Besides, it could not be moved between Slough and Bath, and William had "an inclination to pass some time at Bath which I should not like to do without having an instrument of a large aperture".[12]

And so he decided to make himself a reflector that not only had more 'light gathering power' but was one that he could manoeuvre himself despite his advancing years—and which furthermore was compact enough to be transported between Slough and Bath. He decided that what he needed was a mirror that was 2 feet in diameter but had a focal length of only 10 feet. This mirror would have to be thicker at the edges and more deeply ground in the middle than anything he had previously attempted, but he rightly thought that his skills were now equal to the challenge (Fig. 4.4).

In Bath William arranged for the disc for the new reflector to be cast from Michell's metal. William then ground the disc and polished the resulting mirror, presumably in the house to which Alexander had moved after the death of his wife. William tried the new telescope for the first time in August 1799, before the polishing was complete, and was delighted. With it he could single-handedly "get any object in less than half a minute".[13] But his glory days as an observer were drawing to a close, and eventually, in 1814, he sold the 10-ft to the brother of Napoleon.

Sion Hill proved to be a little too far from the centre of town, and so from midsummer 1800 he took a year's lease on a house in Little Stanhope Street, just around the corner from his old haunts in New King Street. The property—which neither William nor Mary had yet seen—had been empty for some weeks, and their possessions were moved there from Sion Hill without supervision. Much needed to be done to make the house fit for habitation, and as usual Caroline was the answer. William decided that she should uproot herself from the Windsor area and move permanently to Bath, to look after the house there when he was in Slough, and to come to Slough as and when he and Mary went to Bath. Occasionally William and Caroline would have to be in Slough together, when William was minded to have some sweeps with the 20-ft, but this work was drawing to a close.

At the end of June, therefore, a reluctant Caroline took formal leave of her friends at Windsor Castle and moved out of her lodgings, and soon she was on the coach to Bath. With the help of a servant she got William's house into some sort of order, and then she was ready for deskwork to be sent to her, from Slough or from the printers in London.

In mid-November William summoned her to Slough, for he and Mary were planning to spend two or three weeks in Bath. Then, if the weather was suitable for observing, he would return, "that I may have a few sweeps with you before you go

[12] RAS Herschel Archive W.6/8.

[13] RAS Herschel Archive W.2/2.5, f. 52v.

Fig. 4.4 The 10-ft reflector with 24-inch mirrors that William constructed in Bath in the summer of 1799. The previous October, when in Bath, he had written: "Having an inclination to pass some time in Bath which I should not like to do without having an instrument of a large aperture I have it in view to make one of [these] dimensions for the sake of ready transportation and that it may stand on a small place." The "X-ft" (as William termed it) was ideal for the study of individual objects ("I can get any object in less than half a minute"). Although the drawing is not signed, it must be the work of William Watson. Courtesy of the Royal Astronomical Society

back to Bath".[14] But it seems that, once in Little Stanhope Street, William and Mary decided they had made a serious error in taking a year's lease of the house, for they promptly shipped Caroline's effects back to Slough, and never occupied

[14] Mrs J. Herschel (1879), 106.

the house again except for a brief visit by William in the summer of 1801 to wind up his affairs.

This sudden change of plan left Caroline without a home, but fortunately her nephew George Griesbach, eldest of Sophia's five sons all of whom were by now musicians in Queen Charlotte's band at Windsor Castle, had a couple of rooms to spare. But this could be only a temporary arrangement, as it meant that Caroline was two miles from the observatory, an awkward distance in the daytime and an impossible one if she was needed at night. And so after four months she moved to Chalvey, a hamlet only a mile from Slough, where she rented accommodation from a woodcutter.

Her next move, in March 1803, was to lodgings in Upton; and it was not long before rooms became available for her nearby in Mary's old home, Upton House, which Mary was now renting out. After some seven years there Caroline moved into a small house next to the Crown Inn in Slough. It was owned by William and only a stone's throw from the observatory. An ideal solution, one might think, but she was lonely, and it was small consolation that her brother was enjoying a happy family life almost next door. She took a married couple as lodgers, in return for the wife's cooking and cleaning for her. This arrangement lasted until February 1814 when the husband became gravely ill and so his wife could no longer care for Caroline. In addition it seems that the Crown Inn wished to expand into the property. "I became unsettled once more, as the house we occupied was to be let, and I was obliged to move to a small cottage in Slough, at a considerable distance from my brother".[15] There she would remain until she left England after the death of William.

The distance between William's home and Caroline's various lodgings restricted the help she could give William as his amanuensis, for the reference works she needed seemed never to be at hand when she wanted them. More importantly, her peregrinations played havoc with her own career as a comet-hunter, and Caroline the observer achieved nothing of note after her impetuous exit from the Herschel home.

[15] C. A. Lubbock, author's typescript of *The Herschel Chronicle*, chap. 24, p. 26, William Herschel Museum, Bath.

Chapter 5
William's Declining Years

5.1 Light Illuminated

William's major publications in *Philosophical Transactions* relate to the 'construction of the heavens', but half his papers deal with the Sun and the various members the solar system. His second publication was on the mountains on the Moon, and the third was his pioneering investigation into "whether the Earth's diurnal motion is perfectly equable". Saturn had been the very first object he had 'officially' observed, back in 1774, and he returned to the planet time and again over the years, and published no fewer than eight papers reporting his observations.

He had discovered Uranus in 1781, not because he had detected its movement in relation to background stars, but because the superb mirror of his 7-ft reflector had enabled him to see at a glance a suggestion of its planetary disk. Uranus had doubled the scale of the known solar system; but it had not solved the riddle of the surprisingly large gap between Mars and Jupiter. At the turn of the century, a combined attempt was being organised by Continental astronomers to search for the 'missing' planet in the supposed gap, when they were forestalled by Giuseppe Piazzi of Palermo Observatory. At the beginning of January 1801, Piazzi was at work measuring star positions for a catalogue he planned, and he always checked each position by measuring it twice, on successive nights. On 2 January he found to his surprise that a 'star' he had measured the previous night had moved. Piazzi tracked the object for some six weeks before losing it in the glare of the Sun. The problem that then faced the astronomical community was, how to recover the object when it became visible once more, given that Piazzi had observed it for only a fragment of its orbit.

At the end of August 1801, William arrived back home after a month's tour of Wales, to find awaiting him a letter from Johann Elert Bode dated 6 June. Bode informed William of Piazzi's discovery, explained the problem, and asked if William would search for the object in the region where it was thought likely to be found.

M. Hoskin, *William and Caroline Herschel*, SpringerBriefs in Astronomy,
DOI: 10.1007/978-94-007-6875-8_5, © The Author(s) 2014

William was currently engaged in a program of observations with a 10-ft reflector using a magnification of 600, and he decided to use the same instrumentation in a systematic search for Piazzi's 'star'. The problem was, how to examine the region of the ecliptic in an organized fashion, avoiding both gaps and duplications, and William's solution was the 'method of triangles' (and trapezia). He would select a region of the ecliptic defined by three bright stars, and examine the stars (real or supposed) within that triangle, in the hope of finding one that displayed a planetary disk. He would then move on to an adjacent triangle, at whose vertices were two of these three stars plus a different third star, and do the same. And so on.

For weeks William delayed writing to Bode in the hope that the next night would bring success in his search for Ceres, and it was not until 27 October that he replied:

> I have been on the look out for the planet, comet, or moved star, about the place pointed out in the tables you have been so good to send me; but I am pretty well assured that it is no where in that neighbourhood. Were it one thousand times less in bulk than Mars I think it could not have escaped me; I shall however, continue my research, and should I have any success you will immediately hear of it.[1]

In the event, Ceres proved so tiny that even William had extreme difficulty in discerning its disk—and this even after it had been rediscovered at the turn of the year in a position calculated by the mathematician Carl Friedrich Gauss, and William had been told by Maskelyne exactly where to look. His first estimate was that "the real diameter of the new planet, remarkable as it may appear, is less than five eights of the diameter of our moon".[2]

At the end of March 1802 another such object, Pallas, was discovered by an amateur astronomer of Bremen, Heinrich Olbers. William located it and found that it too was tiny, so much so that even little Mercury was over 30,000 times bigger.

It was time for him to go public with his measurements, and for this he needed a name by which to refer to these curious objects, which were planet-like in their movements but not in their sizes. After consultation he decided upon 'asteroid', the name we use today, although he was criticised for this. Piazzi himself preferred 'planetoid', for reasons we might think persuasive, and there were those who held that William had denied the objects the name of planets simply in order that he might remain the only planet-discoverer in history.

In theorizing about the Sun itself, William allowed himself to be guided by considerations outside science, for he was convinced that the Creator had located intelligent beings everywhere throughout the universe: on every satellite (notably the Lunarians), on every planet, and even on every star, including the Sun. To invent the physical conditions that would permit life on the Sun, William proposed that at its core it had a very large, planet-like body on which the inhabitants lived, and that they were shielded from the blazing exterior by clouds.

---

[1] W. Herschel to J. E. Bode, 27 Oct. 1801, RAS Herschel Archive W.1/1.

[2] Dreyer (1912), 1, pp. cix–cxi.

To observe the Sun William used filters, some of them devised by Alexander; and he began to suspect that the heat reaching his eyes varied with the colour of the filter. Careful experiments showed that this was indeed the case, and that the heating increased towards the red end of the visible spectrum. Indeed the heating effect did not end there, for by using a range of thermometers exposed to the spectrum of sunlight William demonstrated the existence of what we term 'infra-red rays' but which he spoke of as "invisible light". William announced this discovery in two papers published in *Philosophical Transactions* in 1800. Sir Joseph Banks wrote to congratulate him: "I hope you will not be affronted when I tell you that highly as I prized the discovery of a new planet, I consider the separation of heat from light as a discovery pregnant with more important additions to science".[3]

The nature of light itself had fascinated William ever since his early days as a member of the Bath Philosophical Society, when he had noted his intention "to examine the arguments for a difference of velocity in different coloured light".[4] Newton had shown that sunlight is compounded from the colours of the rainbow, and builders of refracting telescopes knew all too well that as sunlight passes through a lens the red component is bent the least and the violet the most. This might perhaps be because the particles of differently-coloured light were of different sizes, or of different densities, but it might also be because red light was travelling faster than violet and that the angle through which the lens managed to deflect it was therefore smaller.

Very occasionally a star is eclipsed ('occulted') by a planet. When the planet moves away and so allows the starlight to set off once more on its journey to Earth, all the colours that make up the starlight set off at the same instant. But if the red component is travelling the fastest, then that component will arrive before the rest and for a moment or two the star will appear redder than usual. And the greater the distance of the planet from Earth, the more evident this effect (if it exists) will be.

William's own planet was the outermost of the solar system and the furthest from Earth, and in November 1783 Patrick Wilson of Glasgow University suggested to William that astronomers should therefore keep watch for occasions when his planet passed in front of a star. William was discouraging in his reply. He knew that the reddening effect would be much reduced if the planet were surrounded by an atmosphere; and since he believed the planet was inhabited and that the inhabitants might need to breathe, he was predisposed to expect an atmosphere to be present. He had said much the same when Samuel Vince of Cambridge had written a couple of months earlier, suggesting that the reddening effect might be seen in eclipses of moons of Jupiter, which have the merit of occurring frequently. There was, he told Vince, "an atmosphere surrounding every celestial body[5]" that was capable of causing an eclipse.

---

[3] J. Banks to W. Herschel, 9 April 1800, RAS Herschel Archive W.1/13.B.34.

[4] RAS Herschel Archive W.6/9.

[5] W. Herschel to S. Vince, [Dec. 1783], RAS Herschel Archive W.1/1, 90–91.

Interestingly, he himself made a careful observation of the eclipse of Jupiter's third moon on 28 October 1783. At 5.26 a.m. the moon "was reddish", while 2 min later "it had nearly regained all its light".[6] This was exactly the evidence that Vince and Wilson saw as crucial; but William was serenely convinced that Jupiter has an atmosphere, and so his mind was closed to the implications of what he had seen.

The tests proposed by Vince and Wilson were designed to test a theory; but when Thomas Collinson, a man with links to Bath, stayed the night at Datchet in the spring of 1783, he suggested to William an experiment purely out of his curiosity to see what would happen. For centuries the rainbow-effect created when sunlight is passed through a prism had been a source of interest and amusement, but until the dawning of the era of giant reflectors there had been no way of collecting enough starlight to do the same and view the outcome. William's existing 20-ft had 12-inch mirrors; why not turn it towards Sirius, the brightest of all the stars, and see if the reflector collected enough light to give a visible spectrum?

In fact, William had long been fascinated by the variety of colours to be found among the stars, and especially by what he called "the garnet star", μ Cephei. On 21 May 1783, he had pointed his 10-ft reflector first towards μ Cephei, and then to the much brighter α Cephei nearby. Using his finger and thumb to hold a prism between his eye and the eyepiece, he compared the two spectra. That of α Cephei showed a wide range of colours—red, orange, yellow, green, blue, purple, violet—while that of μ Cephei was more restricted—mainly red, yellow and green.[7] Interesting perhaps, but there was nothing he could do with the information. Decades later, a whole new branch of astronomy—'astrophysics', otherwise known as the New Astronomy—would be developed, to apply to astronomy the lessons that had been learned by physicists when they passed light through a prism in the laboratory. On 21 May 1783, William inaugurated the prehistory of astrophysics.

In April 1798, William's curiosity led him to try again. This time he devised a way of attaching a device to the eyepiece of his 7-ft that held the prism securely while allowing it to rotate. This enabled him to make a careful examination of the spectra of six of the brightest stars. In the remarks that preceded his 1802 catalogue of 500 nebulae, he made an excuse to describe the spectra and how they differed from each other; but again there was nothing he could do with the information.

---

6 RAS Herschel Archive W.3/1.6, 46.

7 For more on William's study of the spectra of starlight, see Hoskin and Dewhirst (2006).

5.2 Companions in Space

William's first major investigation in astronomy had been his search for double stars, in the hope that some if not all doubles were chance alignments, with the two components at very different distances from Earth. If so, observations of a double in which the further star was so remote as to be for practical purposes fixed might reveal the distance from Earth of the nearer star. Michell however, had pointed out that the number of doubles in the sky far exceeded the number to be expected on the hypothesis of chance alignments, and that most must be formed of two companion stars, each in the gravitational grip of the other.

Two decades had since passed, years in which gravity had become in William's eyes the key to the understanding of the construction of the heavens and its evolution over time. He was now much more open to persuasion by Michell's argument, and he decided he would re-examine some of his doubles, to see whether he could find examples where the component stars might indeed have rotated about their common centre of gravity. A mere two observations, separated in time by a couple of decades, could not by themselves demonstrate that the force at work was in fact gravity (under gravity, the orbit of each component would be an ellipse). But changes in the relative positions of the two components might be enough to show that they were indeed companions, bound together by (one or more) attractive forces. And this, in a number of instances, was what William in fact found.

In the case of the famous double star Castor, a lucky chance allowed him to calculate an accurate value for the time the components took to orbit each other. Maskelyne's predecessor as Astronomer Royal, James Bradley, had told him that in 1759 the line joining the components of Castor chanced to be parallel to the line from Castor to Pollux. This supplied William with a third datum, two decades earlier than his own first observation. With this information, he calculated the period of the orbit of the components of Castor to be "342 years and two months", almost identical to the modern value of 350 years.

William made a preliminary announcement of his confirmation of the existence of binary stars in the preliminary remarks to his final catalogue of nebulae, published in 1802. The detailed evidence he supplied in papers that appeared in 1803 and 1804.[8]

In 1802 William was honoured with an audience with Napoleon, during a visit to Paris after peace had been restored between Britain and France. Arriving in the French capital with Mary and John, he first made friends with Simon de La Place, whose theory of the nebular origins of our planetary system chimed with William's theory of the nebular origins of stars. On 30 July the two men breakfasted together. "His Lady received company abed; which to those who are not used to it appears

[8] W. Herschel, Account of the changes that have happened, during the last twenty-five years, in the relative situation of double stars..., *Phil. Trans.*, 93 (1803), 339–82, and "Continuation...", *Phil. Trans.*, 94 (1804), 353–384.

very remarkable".[9] On 8 August he dined in the early evening with the Minister of the Interior, who then took William, La Place, and the Anglo-American physicist Count Rumford to Malmaison. There Napoleon's wife took them round the gardens, and the minister then introduced William and Rumford to Napoleon himself. Napoleon asked William some questions about the construction of the heavens.

> He also addressed himself to M. La Place on the same subject, and held a considerable argument with him in which he differed from that eminent mathematician. The difference was occasioned by an exclamation of the first Consul's, who asked in a tone of exclamation or admiration (when we were speaking of the extent of the sidereal heavens) "and Who is the author of all this!" Monsieur La Place wished to show that a chain of natural causes would account for the construction & preservation of the wonderful system, this the first Consul rather opposed. Much may be said on the subject, by joining the arguments of both we shall be led to "Nature and nature's God".[10]

In the middle years of the decade William fell into bad company, scientifically speaking. He had long been friends with Professor Patrick Wilson of Glasgow University, and when Wilson retired and moved south the two men regularly met for what they termed "philosophical chapters". A subject that fascinated them was the coloured rings ('Newton's rings') that appear when two pieces of curved glass are pressed together. For some time William virtually abandoned astronomy and devoted himself to these rings, and he prepared three papers of excessive length that reported the results of innumerable experiments. His friends were dismayed and the Royal Society most unhappy at the prospect of publishing these interminable papers in *Philosophical Transactions*, but such was William's prestige (to say nothing of his pigheadedness) that the papers were eventually published. When William's son John considered publishing a posthumous collection of his father's papers, he planned to omit these three; and when such a collection did eventually appear in 1912, the editor included them only reluctantly and for the sake of completeness.

One aspect of his work in astronomy that William could not neglect was the upkeep of the 40-ft, which managed to be both a white elephant and a millstone round his neck. After the turn of the century it served almost no scientific purpose, although William at one stage did use it to refresh his memory of some of the Messier nebulae. (By unhappy chance these did not include M51, the Whirlpool Nebula, for when viewing this nebula William might well have realized that it had a spiral structure, and we would now be remembering the 40-ft with the affectionate admiration we reserve instead for Lord Rosse's great reflector at Birr Castle.)

But the 40-ft was the symbol of George III's patronage of science, and the stream of visitors despatched from Windsor to Slough, to admire both the instrument and its royal patron, was endless. There was no question of these visitors actually climbing the frame and peering through the eyepiece, but William

---

[9] RAS Herschel Archive W.7/15, 262.

[10] RAS Herschel Archive W.7/15, 269–270.

(with out-of-season help from Alexander) had to maintain the appearance of a working telescope. Every couple of years or so they would make an effort to repolish the mirror, and for this they had to improvise a crane and, with the help of labourers, hoist the one-ton speculum out of the tube, polish it, and then replace it—all at the cost of immense effort and no little danger.

5.3 A Cosmogony

When William at last gave up his obsession with Newton's rings and returned to the exploration of the construction of the heavens, he drew on his great catalogues of nebulae to present his conception of a universe where change was universal and gravity (or, at least, some such 'clustering power') was the force that brought the changes about. A nebula/cluster that consisted only of nebulosity was in the early stages of development; one that contained both nebulosity and stars was in middle life; and one that was formed entirely of stars was approaching old age. In papers published in 1811 and 1814, he illustrated what he believed to be the successive stages of development by presenting 'articles', each of which was illustrated by examples from his catalogues:

> … and it will be found that those contained in one article, are so closely allied to those in the next, that there is perhaps not so much difference between them, if I may use the comparison, as there would be in an annual description of the human figure, were it given from the birth of a child till he comes to be a man in his prime.[11]

Article 1 of the 1811 paper cited examples of "extensive diffused nebulosity"; by article 33 of the paper the gravitational process had reached the stage of "stellar nebulae nearly approaching to the appearance of stars". The successive articles of the 1814 paper then took the reader "from the first indications of clustering stars, through irregular as well as through more artificially arranged clusters, up to the beautiful globular form".[12] Here we see clearly expressed the modern conception of the life story of a celestial object.

This was the achievement of a musician of minimal education, who was well into his forties before he first became a professional astronomer, and who did so at a time when other astronomers were still in thrall to the clockwork universe of Newton. But it must be admitted that William's final cosmogony is curiously incomplete. In these two papers, he tells us nothing of where the nebulosity of the first article of the 1811 paper has come from—we are told about birth but not about conception. Nor does he say anything about what later happens to the systems of

---

[11] W. Herschel, Observations relating to the construction of the heavens, *Phil. Trans.*, 101 (1811), 269–336, p. 271.

[12] W. Herschel, Astronomical observations relating to the sidereal part of the heavens ..., *Phil. Trans.*, 104 (1814), 248–284, p. 281.

"beautiful globular form" of the closing articles of the 1814 paper—we are told about maturity but not about the death that must follow.

This is curious, for in 1791 he had speculated—plausibly, one might think—that nebulosity began as gatherings of the particles of light given off from the stars that populate the universe:

> How far the light that is perpetually emitted from millions of suns may be concerned in this shining fluid, it might be presumptuous to attempt to determine; but, notwithstanding the unconceivable subtilty of the particles of light, when the number of the emitting bodies is almost infinitely great, and the time of the continual emission indefinitely long, the quantity of emitted particles may well become adequate to the constitution of a shining fluid, or luminous matter... [13]

Again, in his earliest papers on the construction of the heavens he had envisaged globular clusters as ending in gravitational collapse, of which he offered Tycho's nova of 1572 as a possible instance; and this, one might think, was also plausible.

William must surely have been tempted to propose a cosmic cycle in which light particles gathered to provide the nebulosity of the first article, while the globular clusters of the last articles went on to suffer gravitational collapse. The light of the resulting conflagration would presumably supplement the light particles that are constantly being emitted by stars, and this would lead full circle back to the nebulosity of the first article.

Why, then, does he not say as much? We can only guess; but perhaps he thought the successive articles of his two great papers were so close to the observational evidence that they would carry conviction, and he may have been loathe to adulterate this fact-based exposition with speculation. In the papers as published, he had done nothing more than arrange his specimens in a certain order (and suggest that this order corresponded to their life cycle).

To us, William's globular clusters (we continue to use his term) are vast systems that are satellites of the Galaxy itself, typically comprising between fifty thousand and a million individual stars; while his irregular clusters are our galactic (or open) clusters, comparatively modest associations of stars, located within the Galaxy and so usually seen by us as close to the Milky Way. We might say that, for us, William's clusters form two distinct species. But William was unwilling to compromise his vision whereby gravity effects a unidirectional transformation from start to finish; and so he preferred to think that gravity would in time bring about the reformation of an irregular cluster, and mould it sooner or later into the perfection of the globular form.

In the 1780s, William had devoted much effort to investigating the detailed structure of the stratum of stars to which the Sun belongs and which gives rise to the optical effect of the Milky Way. Further experience had convinced him that the two assumptions on which his investigation was based were unjustified, but his belief that the Galaxy is a stratum of stars never wavered. Yet he was never able to

[13] W. Herschel, On nebulous stars, properly so called, *Phil. Trans.*, 81 (1791), 71–88, p. 87.

offer the slightest hint as to how this stratum might have originated. His 1814 paper ended with a sublime reflection on this mystery:

> We may also draw a very important additional conclusion from the gradual dissolution of the milky way; for the state into which the incessant action of the clustering power has brought it at present, is a kind of chronometer that may be used to measure the time of its past and future existence; and although we do not know the rate of going of this mysterious chronometer, it is nevertheless certain, that since the breaking up of the parts of the milky way affords a proof that it cannot last for ever, it equally bears witness that its past duration cannot be admitted to be infinite.[14]

[14] W. Herschel, Astronomical observations relating to the sidereal part of the heavens ..., *Phil. Trans.*, 104 (1814), 248–284, p. 284.

Chapter 6
John's "Sacred Duty"

6.1 The Torch is Handed On

Caroline meanwhile was enjoying robust health, and although she was convinced every year would be her last, she was to live to a great age. She did however, have one serious scare, in the autumn of 1808, when she felt unwell and summoned the Herschel family physician, Dr Pope, to her bedside. For some reason, a week later she dismissed Pope and summoned Dr Phips. But this was a mistake, for Phips wrongly decided she was going blind, and had her kept for a fortnight in a darkened room so that she might practise the skills she would need when the calamity happened.

William, nearly 12 years Caroline's senior, was approaching old age, and his health was beginning to show the effects of endless nights spent at the telescope in the harsh winters then prevalent. He had swept most of the sky visible from Slough, at latitude over 51° north of the equator; but he had been able only to dream of visiting the Cape of Good Hope and viewing the southern skies. Even his attempt to sell a telescope to the King of Naples, and make the excuse to visit Naples and observe that far south, had come to nothing.[1] Worse still, his great catalogues of nebulae and clusters remained the work of a natural historian of the heavens, and in their present form they would be of limited use to future observers who might follow in his footsteps. If his work was to be brought to a completion—thoroughly checked and recast, and then extended to the southern skies—the only hope lay with his son John.

John had been brought up in a hot-house atmosphere, effectively an only child after Mary's son by her first marriage died in his teens. He had been sent to Eton, most famous of the English 'public' schools and close to Slough, but his mother withdrew him after only a few weeks, when she chanced to see her cherished offspring "stripped and Boxing with a great Boy".[2] From then on he attended a

[1] On this see Hoskin (2010b).
[2] Lubbock (1933), 298.

M. Hoskin, *William and Caroline Herschel*, SpringerBriefs in Astronomy,
DOI: 10.1007/978-94-007-6875-8_6, © The Author(s) 2014

small and select school that catered for "the Nobility and Gentry", whose teaching Mary supplemented with private tutors.[3] He was never allowed to play with boys from the nearby Castle, for King George was experiencing periods of 'madness' and under the influence of his son the Court at Windsor was rumoured to be dissolute. In short, poor John was never allowed a proper childhood, especially after his Aunt Caroline withdrew from the immediate family circle.

Things took a turn for the better when at seventeen John entered St John's College, Cambridge. He enjoyed a stellar career amidst a talented generation; as Caroline proudly noted, "from the time he entered the University to his leaving he had gained all the first prizes without exception".[4] John was a true polymath: in 1815 he would miss election to the chair of chemistry by a single vote, and in 1818, friends urged him to apply for the chair of geology. This versatility made his choice of career after his graduation in 1813 a very open matter. He was inclined to follow the example of a friend who had gone into law and who was greatly enjoying it; but he made a half-hearted attempt to take his father's advice, as courtesy demanded.

It was an unfortunate step, for William advised him to take Holy Orders in the Church of England. He could expect a comfortable lifestyle with ample time for his studies, and provided he preached improving sermons he need not worry too much about niceties of theology. This John found shocking—"I cannot help regarding the source of church emolument with an evil eye[5]"—and a rift developed between father and son. A churchman urging his congregation to love their neighbours, William insisted, was to be preferred to the barrister who spends half his time arguing the truth of what he knows to be a tissue of lies.

Mary managed to smooth over the squabble, peace offerings were exchanged, and John spent Christmas 1813 at home with his parents. In the new year, with their grudging acquiescence, he began to study law in London. But the intellectual distractions of the capital proved too great. He became friendly with the chemist William Hyde Wollaston, and more especially with James South, a surgeon who had married into money, given up medicine, and become an amateur astronomer. South had used his wife's fortune to equip himself with precision telescopes. These were well suited to investigations into objects like double stars, and one day he and John would join forces for this very purpose.

These distractions led John to reconsider his plans for a career in law. He was never robust, and in the autumn of 1815, during a period of semi-convalescence, he decided to accept an invitation from his old college to return to Cambridge and coach their undergraduates in mathematics. This very junior position was quite unsuited to someone of John's stellar abilities; but it was a start.

William meanwhile was well into his seventies, and becoming more and more infirm. His worries over the limitations and defects in his astronomical legacy

[3] The source of Mary's wealth is discussed in Hoskin (2010a).

[4] Mrs J. Herschel (1879), 120.

[5] W. Herschel to J. Herschel, 8 Nov. 1813, Herschel Family Archives.

began to dominate his thinking. There was only one hope: his son. And so, in August 1816, William invited John to accompany him on a trip to his favourite resort in Devon.

Although father and son, in age they were more like grandfather and grandson—John was still in his twenties, William not far off eighty. Choosing his moment carefully, William prevailed upon John to sacrifice the congenial career that lay before him in Cambridge, and to become in effect his father's apprentice.

A sacrifice it was. As John wrote to his close friend, the mathematician Charles Babbage:

> I always used to abuse Cambridge as you will know with very little mercy or measure, but, upon my soul, now I am about to leave it, my heart dies within me. I am going, under my father's directions, to take up the series of his observations where he has left them (for he has now pretty well given over regularly observing) and continuing his scrutiny of the heavens with powerful telescopes.[6]

After William's death, John and Caroline were to destroy his rough drafts, to the vexation of later historians, but a sheet of "Work to be done" survives.[7] It appears to be notes for a discussion with John. Among other things, John is to observe the four known asteroids, no doubt to confirm that they were indeed minuscule and formed a species of heavenly body previously unknown. He is to search for undiscovered planets, by sweeping near the ecliptic with the 40-ft facing south (a task his father had embarked on but abandoned after only two hours' viewing), and also by dividing the ecliptic region into triangles with bright stars at the vertices and scrutinising the contents of each triangle in turn. He is to measure the position of β Cygni and make a new determination of its proper motion, for its motion as known in 1783 had been the chief anomaly to William's choice of λ Herculis as the solar apex. He is "To remeasure the angles & distances of double stars" in the hope of proving that the force binding them together was indeed gravity. He is "To finish the Sweeps of the heavens", a task that evidently weighed on William's conscience. And he is "To make a new Sweep of the heavens", to check and revise the catalogues of 2,500 nebulae and clusters that were the fruits of William's sweeps with the 20-ft.[8]

In their present format, the catalogues were of little use either to John or to other observers, for they were arranged as a work of natural history rather than astronomy: by class and then by date of discovery. And the position of each nebula was given not in coordinates but by reference to a nearby star. If John—or any other observer—came across a nebula, it would be next to impossible for him to search the catalogues and see if the nebula was already listed.

To revise and complete his father's great study of the nebulae, John must check that no errors had crept into the descriptions and positions shouted between William and Caroline during their nights of sweeping; fill any gaps; and then

---

[6] J. Herschel to C. Babbage, 10 Oct. 1816, Royal Society HS 2.68.

[7] RAS Herschel Archive W.2/6, f. 25.

[8] William's plans for John are detailed in Hoskin (2012b).

rearrange the objects in a single catalogue designed to meet the needs of other observers. There was no realistic possibility of John's making this check by manoeuvring the 20-ft towards the first entry in one of the classes in a catalogue, and then towards the second entry in some different part of the sky, and so on. If John was "to make a new Sweep of the heavens" and so revise and complete his father's catalogues, he would have to sweep as his father had done, directing the 20-ft to the south and allowing the heavens to rotate before his eyes. But whereas William had had little idea of what would next appear in his field of view, John would need to be equipped in advance with the data on those of William's nebulae he might expect to encounter in the forthcoming sweep, and so be well-placed either to confirm or to correct them. As he did so, he would automatically fill any gaps in William's sweeps.

All the objects in any particular sweep would be at much the same angular distance from the North Pole, give or take a degree or two, and so William's 2,500 nebulae would have to be rearranged, first in 'zones' of North Polar Distance, and then in the order in which they would present themselves to the observer ('right ascension'). To achieve this recast of the existing catalogues would be a massive undertaking, and Caroline was now well into her sixties; but she was equal to the challenge. As the positions of the nebulae in William's catalogues were specified in relation to nearby stars, the first step would be to assemble all these reference stars into zones of North Polar Distance and order of right ascension. The second would be to replace the reference stars by the nebulae themselves.

Within weeks if not days of John's agreeing to become William's astronomical heir, Caroline was on the job. "October, November, December Nothing particular happened … and I took the opportunity of working on my MS. Catalogue at those times when I was left without employment".[9] Within a couple of years, she had completed this first stage of the great undertaking; when she tackled the second stage, William would be dead and Caroline back in her native Hanover.

Perhaps William and John made some effort to recommission the 40-ft, which John was intended to use to sweep near the ecliptic. If so, this quickly brought home to William that the great reflector was beyond redemption as a scientific instrument: "The woodwork is fast decaying and cannot be effectually mended … and I cannot recommend the 40 feet to be kept up." But the 20-ft was crucial to any future plans: "The woodwork of it is also decaying, but the whole apparatus may be renewed at a moderate expense, and the method of repolishing and preserving or restoring the figure of the mirror when tarnished is not attended with such critical difficulties as will occur when the weight of the mirror requires a crane to lift it on and off the polisher" as was the case with the 40-ft.[10]

To repolish the existing 18-in. mirror of the 20-ft would be a significant step towards bringing the reflector back into service. But it was equally important that John the apprentice learn something of his father's hard-won skills in grinding and

[9] Mrs J. Herschel (1879), 126.

[10] Dreyer (1912), 1, p. lv.

then polishing newly-cast mirrors. Killing two birds with one stone, in the summer of 1817 William had a pair of 18-inch discs cast. One John converted into a mirror under the supervision of his father; the other was entirely John's own work.

In his early months as a novice observer, John tackled a number of items from the agenda that William had prepared. On 23 and 24 November 1816, he used the 7-ft to familiarise himself with eleven of Messier's nebulae. In December 1816, he attempted to identify the asteroid Vesta (which he speaks of as a "planet") by calculating its position on the basis of orbital elements due to Gauss; but he failed to detect any moving object in the region and so could make no contribution to the debate over Vesta's size. A few days later, he "began to lay down fields of zodiacal stars with a view to the discovery of new Planets. The instrument used was a night glass, the place swept was a strip of 10° declination on each side of the first point of [Aries] as marked in Bode's Catalogue, and 4° of RA, viz from 0°0′ to 4°0′[11]"; but nothing came of this either. On four nights in March 1817 he made careful observations of Venus.

The orbit of double stars, however, was a topic that appealed to John, who was a gifted mathematician, and many of his observations in the winter of 1816/17 were of doubles. In March 1821, he was to join forces with South in a massive investigation of these orbits. South owned two precision 'equatorials' ideal for the measurement of doubles, and the two friends would each independently use one of these equatorials and afterwards compare their observations. Their results were to be published in *Philosophical Transactions* in 1824 in a paper that ran to no fewer than 412 pages.[12] The following year John bought one of the two equatorials, South emigrated to France, and the two of them thereafter studied doubles independently of each other.

By the spring of 1817, John's initial series of astronomical observations had come to an end. The 20-ft was in good shape (or so it seemed), John had completed his apprenticeship, and he was being distracted by his many other interests, both in science and in the organization of science. Then, in December 1820, James South came to visit him in Slough. South had no experience of viewing with a great reflector, and John agreed to show him the Moon and Saturn through the 20-ft.

South was impressed, and so was John: "I can now believe anything of the effect of reflectors of great aperture." But the woodwork of the 20-ft was in poorer condition than either William (Fig. 6.1) or John had realized, and later in the evening the reflector "took it into its head to fly from its center".[13] Many of the components could be retrieved from the wreckage and re-used, but there was much work to be done, and whether the outcome was a refurbishment of William's 20-ft, or a new instrument that incorporated components from its predecessor, was a moot point. William however was stimulated by the unexpected challenge: "the

[11] RAS Herschel Archive J.1/1, 9.

[12] J.F.W. Herschel and J. South, Observations of the apparent distances and positions of 458 double and triple stars..., *Phil. Trans.*, 114 (1824), 1–412.

[13] J. Herschel to C. Babbage, 19 Dec. 1820, Royal Society HS 2.150.

Fig. 6.1 Portrait in oils of William in 1819 by William Artaud (1763–1823), showing him wearing his insignia as a knight of the Royal Guelphic order to which he was appointed in 1816 by the future George IV, then Prince Regent. The order had been instituted by the Prince Regent a few months earlier, after Hanover became a kingdom following the defeat of Napoleon, and the knighthood was awarded for distinguished services to the United Kingdom or to Hanover. William, as a Hanoverian by birth living in England, was a very suitable appointee, and his son John was to receive the same appointment in 1831. The Guelphic order was however not British but Hanoverian, and William was no longer Hanoverian, having been naturalised by act of parliament. The award, technically foreign, did not entitle a British holder to be called "Sir"; but this point was widely misunderstood in William's day, and William styled himself (and was usually known as) "Sir William". This image is of the copy Caroline commissioned from Artaud for herself (at a price of 16 guineas), and which is now owned by the Royal Astronomical Society

clearness and precision of his directions during its execution, showed a mind unbroken by age, and still capable of turning all the resources of former experience to the best account".[14]

By May 1821 John once more had a 20-ft reflector at his disposal, and the time had come for him to learn to sweep. On the 29th, the eighty-two-year-old William helped his son position a mirror in the tube, and Caroline took her place at the nearby desk for the first time in many years; and with William supervising—surely from the ground rather than the gallery—John made his first sweep. The following night the scenario was repeated, this time with James South in attendance.

[14] J. Herschel to C. Babbage, 22 Jan. 1821, Royal Society HS 2.153.

Caroline solemnly records South's reaction when shown a nebula with two nuclei: "O! Good God! It is worth going to the devil for![15]"

William was to perform one last service to astronomy. Early in 1820, John and his friends met over dinner and decided to found a scientific society dedicated solely to astronomy. The infant society met on 8 February but only briefly, as King George III had died a few days before. They met again on the 29th and elected officers. William was to be one of two Vice-Presidents, John was elected Foreign Secretary, and the Duke of Somerset was chosen as a suitably prestigious first President.

The creation of the new society outraged the aged Joseph Banks, who feared the damage that specialist societies might do to the generalist Royal Society. Banks went so far as to prevail upon Somerset to withdraw as President, within days of his election, and even to resign from the society itself. Various replacements for Somerset were considered, but it was decided to postpone a choice until the end of the year. By this time, Banks had conveniently died. William was now approached and eventually agreed, but on the understanding that in view of his great age his involvement would be purely nominal. He and Caroline culled from their records a modest list of unpublished double stars encountered in sweeps for nebulae, and this served as a token offering to the publications of the infant society.[16]

6.2 Caroline's Return to "Horrible Hannover!"

By the summer of 1822, William was aged eighty-three and slowly losing strength. Caroline had for some time been reflecting on what she should do when her brother died. In all her life, she was to have just one close friend outside the family, Mrs Charlotte Papendiek, who had been a childhood fellow-pupil at a needlework school in Hanover and who had reappeared in Windsor long afterwards as a Lady of the Bedchamber to Queen Charlotte (the post to which Mrs Papendiek's daughter Sophia was also to be appointed a few years later). When the Queen died in 1818, Mrs Papendiek and Sophia had decided to return to their native Hanover. Caroline's brother Alexander, for so long a mainstay both of the musical life of Bath and of the telescope-manufacture in Slough, had retired to Hanover in 1816 when his health began to fail. And Hanover was the home of their youngest brother Dietrich and his extensive family.

She was especially close to Dietrich, to whom the fates had not been kind. From 1808 to 1812 Dietrich had taken refuge in England during the turmoil on the Continent, a refugee who sent home whatever money he could earn from music so as to support his family. His only son had died young, and now a daughter's husband had become insane and aggressive with it. With William on his deathbed,

[15] RAS Herschel Archive J.1/1.

[16] W. Herschel, On the places of 145 new double stars, *Mem. Astron. Soc. London*, 1 (1822–25), 166–181.

Caroline now made over Dietrich her life savings of £500 to help him in his hour of need—and in the hope that her generosity would endear her to his family.

In 1820, Caroline had sent him £30 and asked him to use it to buy her a feather bed, and with that she came close to committing herself to returning to Hanover after the death of William. There was of course an alternative—simply to write off the feather bed, admit her folly (something that never came easily to her), and remain in England. Her grievance against the Mary who had usurped her place in William's heart had long since been replaced by a grudging affection, and they coexisted perfectly happily. John was the apple of his aunt's eye, but he was currently focused on the scientific societies in London, and on his collaboration with South at South's London observatory. Even if John had been observing at Slough, when looking at bright double stars he could easily have made his own notes and kept his own records: John—currently—had no need of a Caroline.

William died on 25 August 1822. To the distraught Caroline there lurked ghosts around every corner of Slough, memories of happy and fulfilling times gone forever. Her cabin underneath the 40-ft was in decay; her desk by the bedroom window near the 20-ft was gathering dust. Hanover beckoned. True, Alexander had passed away the previous year; but Dietrich was waiting for her with his delightful family, and in Mrs Papendiek and her daughter she would once more have confidants to whom she could unburden herself. Living in her childhood haunts in Hanover, she would be able to look forward with anticipation rather than back with melancholy.

And so she sold what furniture she had, packed up her clothes (and the small comet-sweeper, with its happy memories, and the optics of a 7-ft), and prepared for urgent departure. She refused to wait until John had wound up his father's affairs and could accompany her. Dietrich came to Slough to escort her to her new home; it was a kind act, for one of his daughters had recently been widowed and left with a huge family, and in doing what he could to support her Dietrich's own health had suffered.

Caroline's own account of her leave-taking shows that doubts were already entering her mind. Her impetuosity had always been a fatal flaw. The decision years ago to flounce out of William's home and go into lodgings had been a huge mistake; but she could never bring herself to eat humble pie and ask to be re-admitted. Now her decision to return to Hanover was to prove nothing short of disastrous, for a quarter-century still lay before her, during which her life had little purpose and she would move into an embittered old age, alienated from those around her. Even when Dietrich arrived in Slough to fetch her to Hanover, she could still have changed her mind. But she felt powerless in the face of events that she herself had set in train.

She and Dietrich reached Holland after an horrendous sea-crossing, and took the coach for Hanover. To her dismay she found her brother was a changed man, scarred by the recent blows of fate—the death of one son-in-law, the insanity of another. He had become difficult, and resentful of any contradiction. But in Hanover his wife made her very welcome, and she was delighted with the accommodation arranged for her. Great-aunt Caroline was introduced to the

Fig. 6.2 Portrait in oils of Caroline in 1829, by Melchior Gommar Tieleman (1784–1864), court painter to the viceroy of Hanover. Herschel family archives

children, who found the diminutive figure before them anything but great. Mrs Papendiek sent to ask after her within an hour of her arrival.

But from then on life went, slowly but steadily, downhill. Dietrich was forever complaining about his health, but perhaps with justification, for he died just five years after William. Relations between Caroline and his widow then deteriorated. Caroline endlessly suspected the family of having designs on what little money she had, the pension from the Crown and an annuity from William; and in her later years she three times took a piece of paper and bitterly calculated just how much in total she had given Dietrich's family. Mrs Papendiek herself grew infirm, and meetings between them became more and more rare. Caroline (Fig. 6.2) was lonely and felt herself neglected by her much-younger relatives, and yet she knew she could blame only herself. Back in Slough, even if night watches were now beyond her, she could have saved John the daytime deskwork, and helped prepare his papers and books for publication. Worst of all, she knew in her heart that only her inability to admit her blunder was preventing her from returning to England.

6.3 The Catalogues of Nebulae Recast

But one task she did perform for John from her Hanover retreat, and it ranks among her greatest services to astronomy. To revise and complete his father's catalogues of nebulae, John needed them recast into zones of North Polar Distance and order of right ascension, so that he could sweep with the foreknowledge of what he might expect to encounter that evening. In Slough, Caroline had completed the first stage of the task, assembling the reference stars into the appropriate zones. The second stage was to replace the reference stars by the relevant nebulae—easily said, but a massive undertaking. Fortunately, Caroline could carry out this deskwork as easily in Hanover as in Slough.

The task took her well over a year, and for a time her life in Hanover had purpose. John visited her in 1824 while the work was in progress, so that they could agree the format that would best serve his needs. He found her content and with no reason to regret her move to Hanover; but, sadly, that mood would soon evaporate when the work was done.

Caroline's catalogue reached John in April 1825. It runs to 104 pages of numbers. David Brewster, the Scottish physicist, described it as "an extraordinary monument of the unextinguished ardour of a lady of 75 in the cause of abstract science".[17] John was later to write, "I learned fully to appreciate the skill, diligence, and accuracy which only the most boundless devotion could have induced her to undertake, and enabled her to accomplish".[18] As a woman she was not allowed to become a Fellow of the Astronomical Society of London, but she could become a medallist, and in 1828 its Council resolved

> That a Gold Medal of this Society be given to Miss Caroline Herschel for her recent reduction, to January, 1800, of the Nebulae discovered by her illustrious brother, which may be considered as the completion of a series of exertions probably unparalleled either in magnitude or importance in the annals of astronomical labour.[19]

A few years later the Royal Astronomical Society (as it had become) stretched a point and elected Caroline an Honorary Fellow, which delighted her so much that she had this compliment recorded on her tombstone.

Armed with Caroline's catalogue, in 1825 John set to work to sweep the skies for his father's nebulae, "a sacred duty which I cannot postpone to any consideration".[20] In 1833, he published the resulting catalogue of 2,306 nebulae and clusters, and at last William's verified nebulae (along with 525 of John's) were truly accessible to astronomers world-wide in a form they could use and trust. Thirty years later, in his old age, John would bring together all the nebulae so far known, whether the discoveries of a Herschel or of another observer, and publish them in a single General Catalogue; and this would be revised towards the end of the century into the New General Catalogue that astronomers use today. It was Caroline's deskwork, first in Slough and then in Hanover, that enabled this chain of events to be set in motion.

---

[17] Clerke (1895), 132.

[18] Clerke (1895), 132.

[19] Mrs J. Herschel (1879), 225.

[20] J. Herschel to C. Babbage, 12 Feb. 1828, Royal Society HS 2.219.

6.4 "The Completion of My Father's Work"

There remained the challenge of extending William's work to the southern skies that his father had never seen. In 1832, John's mother died, and apart from a handful of minor bequests she left to him her entire fortune. Her death removed an emotional obstacle to his departing on a lengthy expedition to the south, while her money enabled him to go entirely at his own expense and so to plan and execute the work exactly as he saw fit. John (Fig. 6.3) spent four years at the Cape of Good Hope, and he undertook there one of the most intensive observing campaigns that astronomy has known. He returned in 1838 to justified acclaim, and the Queen conferred on him the hereditary title of baronet. He had been one of seven scientists awarded a Guelphic 'knighthood' in 1831, just like his father before him; but because his mother and aunt would expect him to become "Sir" John in consequence—and this was now recognised as incorrect—he had arranged to receive in addition a British knighthood (an honour that has been kept a secret to this day).[21]

Fig. 6.3 Portrait in oils of John Herschel in 1833, by Henry William Pickersgill (1782–1875). By permission of the Master and fellows of St John's College, Cambridge

[21] This is discussed in Hanham and Hoskin (2013).

Soon after his return, he went to Hanover to visit his aunt, taking with him his eldest son, William James. Like many elderly people unused to children, Caroline was convinced the child would come to harm at any moment, and poor William James was allowed nothing to eat unless his father was present and took responsibility. Then the day came when John and his son were to take their leave. Caroline knew this would be the last time she would ever see them, and she prepared a suitable farewell speech—only to discover when she awoke that John had fled, unable to face the trauma of their parting. Caroline managed to forgive him, but with difficulty.

Meanwhile, John had embarked on the task of preparing for publication the mass of observational data he had assembled at the Cape. The volume would be expensive to publish, but John found an aristocratic sponsor. His *Cape Observations*, to give it the brief title by which it is often known, is an astonishing production, a massive tome the like of which astronomy has rarely if ever seen. The title page speaks of it as "the completion of a telescopic survey of the whole surface of the visible heavens commenced in 1825", for John was, and will forever remain, the only human being to examine with a major telescope every square inch of the heavenly sphere (Fig. 6.4).

John had no Caroline to help with the composition and proofreading, and the book made agonizingly slow progress. In March 1844, the month in which she celebrated her ninety-fourth birthday, Caroline wrote to John's wife saying that she feared she would not live to see the publication. At long last, in the summer of 1847, the magnum opus was printed, and John could send Caroline (Fig. 6.5) her copy. When it arrives, he wrote her, "You will then have in your hands the completion of my father's work".[22] The following January, she died.

In her later years, she had grown increasingly regretful of the damage she had inflicted on herself by quitting Slough for "horrible Hannover".[23] But she had become an esteemed resident of what was now a kingdom, and the King of Hanover, and the Crown Prince and Princess, sent their coaches to follow her hearse in an exceptional mark of respect; and by order of the Crown Princess, her coffin was adorned with palm branches.

In the coffin, at her instructions, were placed a lock of William's hair and an almanac that had belonged to her father. She had long since purchased the plot where Isaac had been buried in 1767. Anna had been buried above Isaac 22 years later, and as a result there was no room in the ground itself for Caroline. She had therefore arranged for a vault to be built over the grave, and this is to be seen in the Gartenkirchhof to this day. The visitor can read the inscription, which is a German translation of words that Caroline had written in English. It tells of Caroline's discovery of comets and her share in William's labours, and it records her membership of the Royal Astronomical Society and of the Royal Irish Academy (Fig. 6.6).

[22] Mrs J. Herschel (1879), 342.

[23] C. Herschel to J. Herschel, 4 May 1843, British Library Egerton 3762.

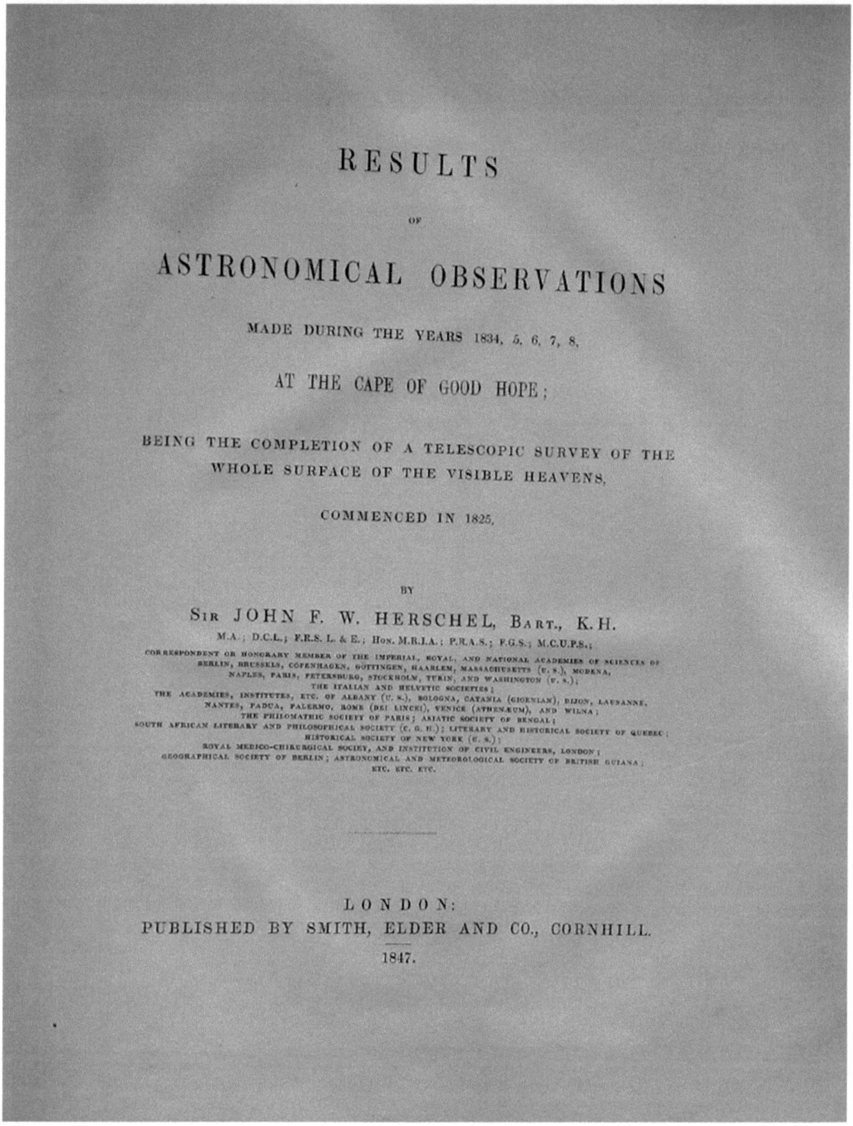
RESULTS

OF

ASTRONOMICAL OBSERVATIONS

MADE DURING THE YEARS 1834, 5, 6, 7, 8,

AT THE CAPE OF GOOD HOPE;

BEING THE COMPLETION OF A TELESCOPIC SURVEY OF THE WHOLE SURFACE OF THE VISIBLE HEAVENS,

COMMENCED IN 1825,

BY

SIR JOHN F. W. HERSCHEL, BART., K.H.

M.A.; D.C.L.; F.R.S. L. & E.; HON. M.R.I.A.; P.R.A.S.; F.G.S.; M.C.U.P.S.;

CORRESPONDENT OR HONORARY MEMBER OF THE IMPERIAL, ROYAL, AND NATIONAL ACADEMIES OF SCIENCES OF BERLIN, BRUSSELS, COPENHAGEN, GOTTINGEN, HAARLEM, MASSACHUSETTS (U. S.), MODENA, NAPLES, PARIS, PETERSBURG, STOCKHOLM, TURIN, AND WASHINGTON (U. S.);
THE ITALIAN AND HELVETIC SOCIETIES;
THE ACADEMIES, INSTITUTES, ETC. OF ALBANY (U. S.), BOLOGNA, CATANIA (GIOENIAN), DIJON, LAUSANNE, NANTES, PADUA, PALERMO, ROME (DEI LINCEI), VENICE (ATHENÆUM), AND WILNA;
THE PHILOMATHIC SOCIETY OF PARIS; ASIATIC SOCIETY OF BENGAL;
SOUTH AFRICAN LITERARY AND PHILOSOPHICAL SOCIETY (C. G. H.); LITERARY AND HISTORICAL SOCIETY OF QUEBEC;
HISTORICAL SOCIETY OF NEW YORK (U. S.);
ROYAL MEDICO-CHIRURGICAL SOCIETY, AND INSTITUTION OF CIVIL ENGINEERS, LONDON;
GEOGRAPHICAL SOCIETY OF BERLIN; ASTRONOMICAL AND METEOROLOGICAL SOCIETY OF BRITISH GUIANA;
ETC. ETC. ETC.

LONDON:
PUBLISHED BY SMITH, ELDER AND CO., CORNHILL.

1847.

Fig. 6.4 "The completion of my father's work": the titlepage of John Herschel's *Cape Observations*. Photograph by Simon Mitton

Fig. 6.5 Engraving of Caroline at the age of 97. In the opinion of the daughter of Caroline's close friend Sophia Beckedorff, it did not "do justice to her fine old countenance. Yet it is extremely like in feature, expression, and deportment"

Fig. 6.6 Caroline's tomb in the Gartenkirchof in Hanover. Her mother and father both lie below her, but the inscription, a translation from the English text she composed herself, makes mention only of her father. Photograph courtesy of Owen Gingerich

It then goes on to acknowledge the presence of Isaac in the ground below. But of Anna, who did her utmost to confine her daughter to a life of domestic servitude, and whose body lies between those of Isaac and Caroline, there is no mention at all. In a final act of revenge, Caroline condemned her mother to an unmarked grave.

Appendix
Visitors' Accounts of the Herschels at Work

We are fortunate that a number of visitors to the Herschel homes in the 1780s have left us detailed accounts of William and Caroline at work in this, their most productive period. These accounts enrich the picture we derive from the siblings' own written records.

The earliest is by the French geologist Barthélemy Faujas de Saint-Fond *(1741–1819), and describes the night he spent at Datchet, 15/16 August 1784, when the sweeps for nebulae had been under way for nearly eight months. From it we learn that already at this date the 20-ft was capable of being moved out of the meridian, and indeed that it could be used to track the rotation of the heavens:*

I arrived at Mr. Herschell's about ten o'clock. I entered by a staircase, into a room which was decorated with maps, instruments of astronomy, and natural philosophy, spheres, celestial globes, and a large harpsichord.

Instead of the master of the house, I observed, in a window at the farther end of the room, a young lady seated at a table, which was surrounded with several lights; she had a large book open before her, a pen in her hand, and directed her attention alternately to the hands of a pendulum-clock, and the index of another instrument placed beside her, the use of which I did not know: she afterwards noted down her observations.

I approached softly on tiptoe, that I might not disturb a labour, which seemed to engage all the attention of her who was engaged in it; and, having got close behind her without being observed, I found that the book she consulted was the Astronomical Atlas of Flamstead, and that, after looking at the indexes of both the instruments, she marked, upon a large manuscript chart, points which appeared to me to indicate stars.

This employment, the hour of the night, the youth of the fair student, and the profound silence which prevailed, interested me greatly. At last she turned round her head, and discovered how much I was afraid to disturb her; she rose suddenly and told me she was very sorry I had not informed her of my arrival, that she was engaged in following and recording the observations of her brother, who expected me, and who, in order that he might not lose the previous opportunity of so fine a night, was then busy in his observatory.

M. Hoskin, *William and Caroline Herschel*, SpringerBriefs in Astronomy,
DOI: 10.1007/978-94-007-6875-8, © The Author(s) 2014

"My brother," said Miss Caroline Herschel, "has been studying these two hours; I do all I can to assist him here. That pendulum marks the time, and this instrument, the index of which communicates by strings with his telescopes, informs me, by signs which we have agreed on, of whatever he observes. I mark upon that large chart the stars which he enumerates, or discovers in particular constellations, or even in the most distant parts of the sky."

This fraternal communication, applied to a sublime but abstruse science, this constancy of study during successive nights, employed in great and difficult observations, afford pleasing examples of the love of knowledge, and are calculated to excite an enthusiasm for the sciences, since they present themselves under an aspect so amiable and so interesting.

Mr. Herschel's observatory, to which I repaired some moments after, is not built on an eminence, nor on the top of a house; he has preferred placing it on a verdant plain, where the wind is not so likely to shake his instruments, and which is sufficiently extensive to permit all the motions such large machines require.

His telescopes are elevated in the air, and mounted in a most simple and ingenious manner: a young man is placed in a kind of chamber below, who, by means of machinery, turns the telescope and the observer together in a circle, with a gradual motion corresponding to that of the earth; thus the reflexion of the star observed is retained on the metallic mirror.

These large machines are, besides, constructed with that precision, solidity, and care, which renders them capable of bearing the intemperateness of the air; and the mirrors are so disposed, that they can be taken out and replaced at pleasure, notwithstanding they are of considerable weight.

Here I saw that ever-memorable telescope with which the eighth planet was discovered. M. Herschel gave to it the name of the King of Great Britain, and called it *Georgium Sidus.* But all astronomers, actuated by a feeling of general gratitude, have, with one unanimous voice, unbaptised it, and given it the name of *the planet of Herschel.*

This telescope, with which I had the pleasure of making observations during two hours, is only seven feet long, and six inches six lines in diameter. Mr. Herschel assured me, that he had made more than one hundred and forty mirrors with his own hands, before he reached that degree of perfection, to which he at last brought them. A telescope of ten feet length is placed beside this one.

This celebrated astronomer has not confined the size of his telescopes to the last measure; there are two others, which are twenty feet long, mounted on large standards rising above the house. The diameter of one of these telescopes is eighteen inches three-fourths, and the mirror weighs one hundred and fifty pounds.

As these superb instruments are of the Newtonian kind, which requires the observer to be beside the object-glass, Mr. Herschel has constructed an apparatus of ingenious mechanism, by which the farther end of the telescope can be reached with ease and safety. There the observer finds a turning chair so disposed, as to enable him to sit at his ease, and follow the course of the stars. A domestic, placed below the telescope, by means of an ingenious combination, moves it softly and gradually along with the observer, and all the apparatus.

Thus William Herschel has been enabled to discover, distinctly, those innumerable stars, which form the most pale and distant parts of the galaxy.

With these instruments he has been enabled to observe that multitude of double stars, as well as so many nebulae, with respect to which astronomers had only vague and uncertain notions; with them, too, he has undertaken to count the stars of the sky, and has made most astonishing discoveries.

Placed at the upper end of his telescope, when the indefatigable astronomer discovers in the most deserted parts of the sky a nebula, or a star of the least magnitude, invisible to the naked eye, he informs his sister of it, by means of a string which communicates with the room where she sits; upon the signal being given, the sister opens the window, and the brother asks her whatever information he wants. Miss Caroline Herschel, after consulting the manuscript tables before her, replies, brother, search near the star *Gamma*, near *Orion*, or any other constellation which she has occasion to name. She then shuts the window, and returns to her employment.

That man must be born with a very great indifference for the sciences, who is not affected by this delightful accord, and who does not feel a desire that the same harmony should reign among all those who have the happiness to cultivate them. How much more rapid would their progress then be!

We commenced our observations with the *Milky-way.*

The telescope of twenty feet discovered to us, in the palest and most distant part of the heavens, an immense number of bright stars, quite distinct and separate from each other.

Mr. Herschel then directed the instrument to the star in the foot of the Goat, which emitted so strong a light as to affect the eye. On making its light fall upon a paper written in very small characters, we could discern and count the lines with ease. It is curious thus to distinguish objects by the glimmering of a star, that is, a sun many hundreds of millions of miles removed from the confines of our system.

The double stars, which are not visible with the most powerful acromatic glasses, appear separate and very distinct, when viewed with the telescope of twenty feet long.

Mr. Herschel made me observe the nebulae of M. Messier, at first with the telescope of seven feet, that is, the one which served to discover the planet. These little specks appear still nebulous with that instrument; and one perceives only a feeble and obscure glimmering. But the telescope of twenty feet permits one no longer to doubt that they are clusters of stars, which appear confused only because of their immense distance; by this telescope they are found to be perfectly distinct.

Mr. Herschel requested me to direct my principal attention to the stars which he was the first to discover to be of different colours from each other, and among which some are seen which border on blue, others on orange, and several on a bluish colour, &c.

It is certainly neither to an optical illusion, nor to the effect of the mirrors and lenses, which Mr. Herschel uses, that we ought to attribute this difference of colour. I started every possible objection upon the subject; but the learned observator always answered them by facts, to which it would be unreasonable to reply.

Thus, for example, he repeatedly directed the telescope to two double stars of pretty nearly the same magnitude, and separated from each other by a small interval only; that is, small in appearance, for the interval must be immense if we consider their distance from earth. Both were of the same colour, and emitted the common white light of the stars.

On directing the same telescope immediately after to other double stars near them, the one appeared to be evidently blue, and the other of a silver colour. The blue star was in some instances on the right, and in others on the left. I saw also some single stars of a blue appearance, several of a bluish white, and others of an orange colour.

Mr. Herschel said to me with much modesty, that this discovery was not of very great merit, since it was easy to make it without recurring to large telescopes; acromatic ones with large object-glasses being sufficient to discover the coloured stars above mentioned.

The observations, however, of Mr. Herschel were at first much disputed, for it is much easier to deny than to examine. But they were soon confirmed, as they deserved to be, by the astronomers of Germany and Italy, and by M.M. Cassini, Mechain etc. of the observatory of Paris.

Mr. Herschel shewed me a pretty large work on the stars; which he designs to publish as soon as it is brought to a conclusion. He has confirmed what has been long since observed, that several stars distinctly marked in the ancient catalogues, and of which some are even laid down in the celestial Atlas of Flamsteed, have entirely disappeared. It is thus probable, that there sometimes happen great revolutions and terrible catastrophes in several parts of the system of the universe; since, if the stars were suns, their extinction must have annihilated the organised beings who existed on the planets which they illuminated.

Jupiter, viewed with the telescope of twenty feet, appears much larger than the full moon. His parallel belts are very distinct, and his satellites are of a truly astonishing magnitude.

On directing the same telescope towards Saturn, we saw his ring in the most distinct manner, and also the shadow which it projected on the body of that immense and singular planet. Mr. Herschel shewed me the sky, and even several stars, in the interval between the moveable ring and the planet. By means of some luminous points which are remarked in the ring, he was enabled to discover that this solid circle has a rotation from west to east in the same manner with the other planets of our system.

The micrometer which Mr. Herschel uses is composed simply of two threads of silk, very fine, well stretched and parallel, which may be moved to a greater or shorter distance at pleasure. The instrument of parallel threads was known before, but this acute observator has perfected it, by finding an easy method of turning one thread over the other at pleasure; so that, on placing them in the telescope, he can take angles with the minutest precision.

The inventor of such large telescopes is far from having confined himself to those of twenty feet long. He was engaged in making the necessary preparations, to construct one of forty feet in length, and of a proportionable diameter.

Mr. Herschel's intention, in constructing telescopes of this great size, is not so much to magnify the object, as to obtain, by the aid of mirrors of such a vast field, a more considerable quantity of light. This project is new and excellent. He told me, that he expected to encounter great difficulties in carrying to perfection a telescope of that great dimension and weight; but that he, at the same time, expected such great effects from it, that nothing should be capable of discouraging his progress.

I remained until daylight in that astonishing observatory, constantly occupied in travelling in the heavens, with a guide, whose boundless complaisance was never wearied by my ignorance, and the importunity of my questions. I passed about seven hours there, employed without intermission in observing the stars. It was impossible to think the time long, when spent in an employment of so profitable, and, with respect to me, curious information. That delightful night appeared no more than a dream to me, and seemed to last only a few instants; but the remembrance of it is indelible; and the grateful recollection of the kindness with which Mr. Herschel, and his interesting sister, condescended to receive me, will never be erased from my heart.

I left *Datchet* (the name of the place where Mr. Herschel resides) about eight in the morning....

From Barthélemy Faujas de Saint-Fond, *Travels Through England, Scotland, and the Hebrides*, 1 (London, 1799), 63–77. Translated there from *idem*, *Voyage en Angleterre, en Écosse et aux Îles Hébrides*, 1 (Paris, 1797), 74–90. Based by Saint-Fond on his manuscript transcribed in "Une visite chez William Herschel", by Audouin Dollfus, *L'astronomie*, ci (1987), 135–143.

A few weeks later, on 6 January 1785, JEAN-HYACINTHE DE MAGELLAN (João Jacinto de Magalhães, 1722–1790), a Portugese resident in London and Fellow of the Royal Society, visited Datchet:

I stayed for the night of 6/7 January last at Mr Herschel's, near Windsor, in the village of Datchet, and had the luck of having a clear night. His large 20-ft Newtonian telescope is located in his garden under the sky with a very simple and convenient mounting. A servant standing below turns a crank leaver alternatively forwards and backwards, until a hammer strikes as soon as the telescope has been raised or lowered by the width of the field of view. This motion is transmitted by a wire into a neighbouring room, and a hand is turned on a dial, whose scale correspond to the angle of elevation of the telescope, calculated in a table. Next to this instrument stands a pendulum clock, adjusted to sidereal time and showing right ascension.

In this room sits the sister of Mr Herschel and she has Flamsteed's sky maps in front of her. When he gives a signal, she notes in a log the declination and right ascension, and also writes down the other circumstances of the appearance. Mr Herschel is investigating the whole sky in this way, without so much as missing a single part of it. He usually observes with 150× magnification (in diameter) and is confident that he will have surveyed everything that is above the horizon within four or five years. He showed me a book in which all his observations so far are recorded, and I was astonished by the amount of what he has already investigated in the sky. With his telescope he examines each time 2 deg 15 min in declination, and he lets each single star pass at least three times through the field, so that it is impossible that anything could escape him. He already has found nearly 900 double stars and the same number of nebulae.

I went to bed one hour after midnight and by then he had so far discovered that night four or five new nebulae. The thermometer in the garden showed 13° Fahrenheit, but unaffected by this Mr Herschel observed during the whole night, except that he has a rest of a few minutes every three or four hours, when he strides up and down in the room I mentioned.

His sister is like him, fascinated by astronomy, and she has considerable knowledge of the calculations involved. For some years now Mr H has missed no opportunity to observe the sky, if the weather allows it, and this always in the open air, for he is convinced that the telescopes work well only at a constant temperature equal to that of the air. He tries to protect himself with clothing against the rough weather conditions, is fortunately blessed with a robust constitution, and thinks in this world of nothing other than the heavenly objects.

He has promised to me in a very friendly way, to have built the telescopes devised by him that have been ordered by me for European observatories, under his supervision and only for the service of astronomy and without personal interest; he also wants personally to give the last finishing touch to the production of the mirrors. A telescope of 7-ft focal length with all the optical accessories and micrometers costs around 200 guineas. They are very lightweight and can be moved by a single person with all their mountings. The 10-ft telescopes, of which he is currently constructing four for the king, cost about 600 guineas, and one of 20 feet, with all the mountings, will be between 2,500 and 3,000 guineas.

I would like you to make this public for the benefit of astronomy.

From *Astronomisches Jahrbuch für das Jahr 1788* (Berlin, 1785), 162–164, transl. by Thomas Maschberger.

John Smeaton (1724–1792), the civil engineer remembered today for building the Eddystone Lighthouse, visited William late in 1785, and on 4 November he sent a report of his visit to William's arch-rival in the building of great reflectors, Rev. John Michell of Thornhill, Yorkshire, formerly Woodwardian Professsor of Geology at Cambridge:

My dear Friend

Since my last I have been to pay my Visit to Mr Herschall, and according to my promise proceed to give you some scantlings of what I have seen; and indeed he has so much originality about him, as well as natural ingenuity, accompanied with great readiness & dexterity, that to enter into the detail would be far to exceed the bounds of a Letter; I will therefore first enter into the great outlines, & fill up as I can.

You must know, till this visit, I have held the Doctrine about Telescopes, that I believe is the common one; that having fixed upon a proportion, that you by experience find to do well, in any one species of Telescope, that what you are to expect from any other size of the same species, is in proportion to the squared root of the length; so that increasing the length four times, your Telescope will allow you to take an image of double the Diameter; every point of it, being illuminated with the same quantity of Light, & painted with an equal degree of distinctness & precision. This Idea & Expectation I carried with me to Thornhill; & carried the same to Clayhall [the Herschel home]; but I did not bring it back with me: Mr Herschalls doctrine will illustrated his pursuits, better than minute descriptions.

Whatever his doctrine originally was, experience has taught him that *large Surfaces* of Speculums, are not to be ground and polished so as to preserve so accurate a figure, as those of a small or moderate size; he therefore divides the maximum that Telescopes may be expected to bring out, into three distinct Classes: 1st. the greatest possible degree of magnifying power, where there is a sufficiency of Light; 2nd. by the greatest degree of distinctness where there is also a sufficiency of Light; but from the natural size of the object, does not require the greatest degree of magnifying power; and 3rdly the greatest degree of Light where the Objects are naturally obscure; which will afford discoveries, which cannot be brought out either by great degrees of magnifying power, or Capacity of distinctness, where, on these accounts, a sufficiency of Light is wanting.

In Conformity with this doctrine; his principal discoveries have been made upon the Stars, where the greatest degree of magnifying powers have been required, and used, with his original Telescope of 7 feet focal length, which he has pushed to between six & 7,000 times: the greatest discoveries have also been made with *these* where the greatest distinctness has been required, under a moderate degree of magnifying power (the Diameter of the Speculum of this Telescope being no more than 6½ inches) and also for the same purpose he finds his 10 feet Telescope applicable, the Diameter being nine inches. But for objects naturally obscure, he can distinctly see that with his Twenty feet Telescope (the Diameter Nineteen Inches, and which is seldom charged with Magnifying power of more than Two hundred times) which the other will not reach.

With this Telescope, he is now and has been for some time past at work, as he calls it, *sweeping* the Heavens. The whole Apparatus can upon occasion be turned to any Azimuth; but is chiefly used, the Telescope turning in the plain of the Meridian. The butt end, or great Speculum end of the Tube, is supported immediately upon the Ground; the other end of the Tube, is raised & lowered by a Tackle, supported at top upon a double equilateral Triangle (or thereabouts). The Observer is also hoisted up in a Chair, that rocks on Rollers upon the inclined Legs of the Triangle next the Eye Glass; and the Eye Glass is brought to answer to this straight line, by sliding the butt of the Telescope nearer the Centre of the whole Machine; and by the same means it can be put into a vertical Position. The raising the Chair and the sliding of the Butt are done by separate Tackles respectively; touched only occasionally; but the main Tackle that raises the Telescope when brought to its intended Elevation, that is polar distance, is worked by a distinct motion; that causes it to rise and fall alternately, through a space of 2 degrees of the Meridian; which being done with some degree of briskness; a fillet in the heavens is Examined at once of 2 degrees broad; the motion of the heavens in AR bringing on the Objects in Succession.

By way of Register, large sheets of Paper are prepared, marked & numbered; being ruled into parallel lines, cross and cross, and about ¼ inch distance; a small square of this kind representing ¼ of a degree in AR and declination. All those that are examined are marked with a cross, and those that have been seen, but not fully examined, with a stroke one way; and when afterwards seen to satisfaction, the cross is completed; the place and species of the object is also marked upon the Paper. In this Operation three persons are Concerned; a Labourer works continually the handle backwards & forwards, for performing the destined range; and in this he is prevented from ranging too little, or too much, by a small piece of Machinery that strikes a Bell, at each end of the Range; he also stops on notice; and if any thing comes requiring this notice, and the Object to be pursued, the telescope can, by an Apparatus which occasionally heaves it from its meridian bearing, pursue it in right Ascension for near a quarter of an hour. And that there may be no need for the Observers Eye to be taken from the Eye Glass; an Assistant (Mr Herschall's Sister) situate in an adjacent Room with the squared sheet before her, upon it notes down, and in a book writes down what is dictated. The Time she has from the Clock before her; and the polar distance by a piece of Machinery, which continually shews the degree and minutes; and is worked by a string actuated by the Telescope, in rising and falling; comes into the room, & winding round a small Barrel, performs the requisite Motions. The Telescope is set to its altitude by a small Quadrant, fixed upon it; & the Corresponding index is regulated answerable to the stretching of the cord of communication, by observing the first known Star that passes of Flamsteeds Catalogue. By this means what has been done, & what is to do, is distinctly seen by the Sheets. In this way many hundred of Nebulas have been discovered, not only unknown before, but wch no ordinary Telescopes will reach.

The speculum of the great Telescope of 40 feet, is cast, but was not got home; it is 4-ft Diameter and about 1,050 lb wt. Mr. Hirshall tells me, there is a Warehouse in Thames Street, where they keep for Sale, metal ready made into Ingotts, of which they have two sorts; what they call *White* Metal, and *Bell* Metal; I suppose such as the Bells of Clocks are made of; but he did not know exactly their Composition: for his Speculum, they put two Ingotts of Bell Metal to one of white Metal. He thinks it a lower metal than what he used for his former Specula of 19 inches, vizt 17 oz ¾ of Tin to 20 oz of Copper (I am not sure however whether I remember right, but you probably will guess). He does not propose it to magnify more than the present ones of 12 inches, but to take the whole advantage in Light. He makes his Specula flatt upon the backside; the thickness of this last great one at the edge, was to have been two inches; but by some shrinking in the mold; & particularly in the middle, I understand it is not there above 1 1/3 Inch; & also less at the edge than it was to have been; so that it is hollow on the back as well as face; but as it is come pretty well upon the face, he proposes to make use of it; & when he has got this rigged out, he proposes to cast another; having duplicates of all he makes; so that while one is in use, the other can go to the polisher. They are made to be inclosed in brass boxes; and their weight lays simply upon several thicknesses of Cloth. They are polished in their Boxes; and are made to go in & take out so conveniently, that they are very frequently put into their Tubes, and tried with an Object, while under the operation of polishing; & to those frequent Trials, he ascribes the principal Cause of his success in these operations.

From what I have said, instead of looking at the Satellites of Jupiter, and expecting them to form one Lucid point, or round Surface, I would wish you to turn your great Telescope to some of the *Nebulae*, for, as the surface of your Speculum, is more than double any Mr Hirshall *yet* has, it is possible you may make some discoveries as yet unlooked for….

Ever yours
J. Smeaton

*NB the Diamr of Mr Mitchells Speculum is 29½ inches; & focal Length 10-ft.

From Royal Astronomical Society MS Radcliffe Hornsby 78.

The novelist Fanny Burney *(1752–1840) was the daughter of Dr Charles Burney, a would-be astronomical poet and friend of William who tried William's patience by reading him his interminable verses. Fanny's diary contains a number of entries recording occasions when her path crossed that of the Herschels:*

[August 1786] In the evening Mr. Herschel came to tea. I had once seen that very extraordinary man at Mrs. de Luc's, but was happy to see him again, for he has not more fame to awaken curiosity, than sense and modesty to gratify it. He is perfectly unassuming, yet openly happy; and happy in the success of those studies which would render a mind less excellently formed presumptuous and arrogant. The King has not a happier subject than this man, who owes wholly to His Majesty that he is not wretched: for such was his eagerness to quit all other pursuits to follow astronomy solely, that he was in danger of ruin, when his talents, and great and uncommon genius, attracted the King's patronage. He has now not only his pension, which gives him the felicity of devoting all his time to his darling study, but he is indulged in licence from the King to make a telescope according to his new ideas and discoveries, that is to have no cost spared in its construction, and is wholly to be paid for by His Majesty.

This seems to have made him happier even than the pension, as it enables him to put in execution all his wonderful projects, from which his expectations of future discoveries are so sanguine as to make his present existence a state of almost perfect enjoyment. Mr. Locke himself would be quite charmed with him. He seems a man without a wish that has its object in the terrestrial globe.

At night, Mr. Herschel, by the King's command, came to exhibit to His Majesty and the Royal Family the new comet lately discovered by his sister, Miss Herschel; and while I was playing at piquet with Mrs. Schwellenberg, the Princess Augusta came into the room, and asked her if she chose to go into the garden and look at it. She declined the offer, and the Princess then made it to me. I was glad to accept it, for all sorts of reasons.

We found him at his telescope, and I mounted some steps to look through it. The comet was very small, and had nothing grand or striking in its appearance; but it is the first lady's comet, and I was very desirous to see it. Mr. Herschel then showed me some of his new-discovered universes, with all the good humour with which he would have taken the same trouble for a brother or a sister astronomer: there is no possibility of admiring his genius more than his gentleness.

December 30 [1786]. This morning my dear father carried me to Dr. Herschel. That great and very extraordinary man received us with almost open arms....

At that time of day there was nothing to see but his instruments: those, however, are curiosities sufficient. His immense new telescope, the largest ever constructed, will still, I fear, require a year or two more for finishing, but I hope it will then reward his labour and ingenuity by the new views of the heavenly bodies, and their motions, which he flatters himself will be procured by it. Already, with that he has now in use, he has discovered fifteen hundred universes! How many more he may find who can conjecture? The moon, too, which seems his favourite object, has already afforded him two volcanoes; and his own planet, the Georgium Sidus, has now shown two satellites. From such a man what may not astronomy expect, when an instrument superior in magnitude to any ever yet made, and constructed wholly by himself or under his own eye, is the vehicle of his observation?

I wished very much to have seen his sister, whose knowledge in his own science is so extraordinary, and who herself was the first discoverer of the last comet; but she all night, and was then in bed.

Mr. Smelt joined us, by appointment; and the Bishop of Worcester came afterwards, with Dr. Douglas, to whom I was then introduced....

By the invitation of Mr. Herschel, I now took a walk which will sound to you rather strange: it was through his telescope! and it held me quite upright, and without the least inconvenience; so would it have done had I been dressed in feathers and a bell hoop—such is its circumference. Mr. Smelt led the way, walking also upright; and my father followed. After we were gone, the Bishop and Dr. Douglas were tempted, for its oddity, to make the same promenade.

[1787] On the evening they left me, my kind Mrs. Delany carried me to Dr. Herschel's.... Dr. Herschel is a delightful man; so unassuming, with his great knowledge, so willing to dispense it to the ignorant, and so cheerful and easy in his general manners, that were he no genius it would be impossible not to remark him as a pleasing and sensible man.

I was equally pleased with his sister, whom I had wished to see very much, for her great celebrity in her brother's science. She is very little, very gentle, very modest, and very ingenuous; and her manners are those of a person unhackneyed and unawed by the world, yet desirous to meet and to return its smiles. I love not the philosophy that braves it. This brother and sister seem gratified with its favour, at the same time that their own pursuit is all-sufficient to them without it.

I inquired of Miss Herschel if she was still comet-hunting, or content now with the moon. The brother answered that he had the charge of the moon, but he left to his sister to sweep the heavens for comets.

Their manner of working together is most ingenious and curious. While he makes his observations without-doors, he has a method of communicating them to his sister so immediately, that she can instantly commit them to paper, with the precise moment in which they are made. But this means he loses not a minute, when there is anything particularly worth observing, by writing it down, but can still proceed, yet still have his accounts and calculations exact. The methods he has contributed to facilitate this commerce I have not the terms to explain, though his simple manner of showing them made me fully, at the time, comprehend them.

The night, unfortunately, was dark, and I could not see the moon with the famous new telescope. I mean not the great telescope through which I had taken a walk, for that is still incomplete, but another of uncommon powers. I saw Saturn, however, and his satellites, very distinctly, and their appearance was very beautiful.

October 8 [1788]. We returned to Windsor at noon. Mrs. de Luc sent me a most pressing invitation to tea, and to hear a little music.... Dr. Herschel was there, and accompanied them very sweetly on the violin: his new married wife was with him, and his sister. His wife seems good-natured: and she was rich too! and astronomers are as able as other men to discern that gold can glitter as well as stars.

From *Diaries & Letters of Madame D'Arblay*, ed. by Charlotte Barrett (London, 1905).

In September 1786 the Herschels were visited by the German novelist Sophie von la Roche *(Marie Sophie Guntermann, 1730–1807), who was on a tour of London and the home counties:*

Thursday, 14 Sept. 1786 ...We spoke of the famous astronomer, Herschel, and Mr. [Carl Godfried] Woide told me about this great man's sister, who accompanies him on his path to glorious immortality, for not only does she help him in his calculations, but in his absence recently discovered a new comet, and enjoys a claim to her brother's great reputation in the matter of the telescope discovered and perfected by him, for when in order to finish it he was forced to remain at his polishing-wheel thirty hours without a break, she assisted him and kept him company, fed him with the food and drink he required, and wiped the sweat from his brow. They live in a small cottage near Windsor, and I hope to make their acquaintance....

Monday, 18 Sept. 1786 ... Crossing some very pleasant country, we very soon came to Slough, where Mr. Herschel and his estimable sister inhabit a solitary, simple dwelling. I was overcome by an intense feeling of respect on seeing this brother and sister—Mr. Herschel full of that humour and philanthropy befitting a noble, sapient man, his sister all gentleness, sensibility and humility; both through their close contact with the constellations are raised above all artificiality and conceit. On a small portion of meadowland there stands a simple wooden scaffolding where Mr. Herschel made the new discoveries which interested the entire learned world, and found the greatest of all the stars, which he named [the Georgian Star]. Even the mechanism of this structure is witness to Herschel's tender love for this science. Upon one side is fixed an enclosed chair, from which

Herschel makes observations, and which can be raised without his speaking a word or interrupting his observations, by a man sitting on the other side in a spacious box-like contrivance well protected against wind and weather, on Herschel's ringing a bell when he wants to be raised higher, which also acts as a signal when he is high enough. I seated myself in this chair with true reverence; the dear man raised it for me; I could see over his garden and the surrounding neighbourhood, beheld the heavens and wished I could be here during a bright starry night....

As my glance fell on the broad horizon which Herschel's eye penetrates through a thousand suns, I bade the heavens protect the brother and sister.

At my request Mlle. Herschel picked me a few daisies which were growing at the foot of this structure, and were all the more valuable for having chosen such a spot and such a hand to pluck them; for this hand takes notes and calculations, as from this simple erection her great brother observes the most wonderful of God's inanimate creatures.

Inside I saw how his chamber communicates with the scaffolding, so that Herschel can move the pointers to two disks, in front of which his sister sits with astronomical charts before her, and notes which portion is being observed.—Can there be a finer mental communion between brothers and sisters than this?

On dull days—a frequent occurrence with the English firmament, since the island is so placed as to afford free entry to mists and vagrant breezes—they prepare their observations for press; read and review the works appearing on astronomy, and entertain one another with music, in which both excel, and seem to believe in the harmony of all creation, although that of the higher spheres is quite inaudible. For me this room became a temple, with a garden leading into it for portico.

I then sat down to the telescope through which Mlle. Herschel had this spring discovered a comet; for this noble creature continued the astronomical researches during her brother's trip to [Hanover], on a visit to his mother, so that science should lose nothing by the fulfilment of filial affections. I peeped through this telescope with real sympathy....

We were then conducted into a hut specially built for the large catoptric telescope designed and executed by Mr. Herschel, 40-ft long and 5-ft in diameter, which is to take a mirror weighing 1,000 pounds, cast and polished by him.

My dear friend and I had the pleasure of seeing this worthy man approach us earthlings, through this remarkable telescope while it lies earth-bound, for in a couple of months his spirit will soar with it to the utmost heights, in contemplation of a thousand new worlds and their planets....

From this great man who, by his spirit and invention of excellent instruments, has perfected a science which has been the source of uninterrupted study during five millenniums, we went straight to Mr. Jervais, the discoverer of a lost art, to which he gave new life and vigour; to wit, that of staining glass....

From Sophie in London 1786, being the diary of Sophie v. la Roche*, transl. from the German (London, 1933), pp. 159–160, 191–193.*

In July 1787, Marc-Auguste Pictet (1752–1825), Professor of Philosophy at Geneva, visited the Herschels and later that year contributed to the Journal de Genève *a lengthy account of his experiences. The length is due in part to the fact that few of his readers were astronomers and so he needed to explain such matters as the difference between reflectors and refractors. Pictet saw the 40-ft under construction, and Herschel explained to him how the 20-ft was currently used for sweeping; but when writing his account from memory Pictet sometimes confused the two instruments:*

The frequent questions put to me concerning Herschel's large telescopes, which I saw recently in England, made me think that the relevant information would be of interest to your readers….

Mr Herschel himself is as remarkable as his achievements and arouses even more curiosity. Nature has granted him a happy organisation, a lively imagination, a good head, sound health, and the ability to persevere to the limit—in other words, he was made to be an astronomer…. Miss Caroline Herschel, his sister, shares her brother's passion for this beautiful science. She has a grasp of its theory that is unique for a woman. She has also the talent for observation, and there has been a recent announcement of her discovery of a comet at a time when her brother was absent. They live together in a house in the country some twenty miles from London and two miles north of Windsor. There they enjoy a life of great contentment. The love of science and the hope of making discoveries inspire their work, and the encouragement they offer each other eases the fatigue inseparable from the way of life they have chosen….

The two large telescopes of Herschel, namely the 20-ft of which I have spoken and the 40-ft, are placed side by side in the open air on a lawn in front of his house, and their constructions are very similar. The mirror of the 40-ft is 4 feet 7 inches in diameter and weighs 1,035 pounds…. Besides the most perfect polish, the mirror of the telescope must have a parabolic figure. The work, as directed by Herschel, to achieve these two conditions, would make a subject for a picture. In the middle of his workshop there rises a sort of altar: a massive structure terminating in a convex surface on which the mirror to be polished is to rest and to be figured by rubbing. To do this the mirror is encased in a sort of twelve-sided frame, out of which protrude as many handles which are held by twelve men. These sides are numbered and the men who are stationed at them carry the same numbers on the strong linen overalls which protect their clothes from the splashes of the liquid, which from time to time is introduced between the mirror and the mould to give the polish.

The mirror is moved slowly on the mould, for several hours at a time and in certain directions, and pressure more on certain parts of the surface than on others tends to produce the parabolic figure. It is then removed on a truck and carried to the tube, into which it is lowered by a machine expressly contrived for the purpose. This labour is repeated every day for a considerable time and by the observations he makes at night, Herschel judges how nearly the mirror is approaching the standard he desires….

The tube of the telescope at the base of which is placed the mirror, is 4-ft 10 inches in diameter and 40-ft in length.... This massive machine must be capable of the same movements as an ordinary telescope, and Herschel shows himself to be as fine a mechanician as he is an astronomer by the way in which he has achieved this at the first attempt.... The tube is suspended at ¾ of its length by a large rope, which passes over certain pulleys and above all over a pulley attached to an overhead beam, from which it descends again to pass around a capstan at the foot of the machine. This combination multiplies the force to such an extent that Miss Herschel can cause the telescope to rise and fall all by herself....

The telescope is capable of two movements, one in the direction parallel to the horizon, and the other perpendicular to this. Each movement is carried out in a different manner, according to whether the movement is to be large or small. If the telescope is to be pointed towards a certain point in the sky, to the Pole Star for example, the entire structure moves through a semicircle on the supporting walls, around the anchor point in the centre. In this way the telescope passes from the south, its usual direction, to the north. The upper end of the tube is then raised by a crank applied to the capstan, to the altitude of the Pole Star.

But in normal use there are no such big movements: it is almost always facing south, in the plane of the meridian....

The constant—and almost unique—purpose of the investigations of Mr Herschel is to explore or *sweep* (as he vigorously expresses it in English) all the celestial regions visible at our latitudes, with sufficient care to ensure nothing escapes him....

But it is not enough merely to observe; one must make records of the observations made and the regions examined, to avoid the risk of sweeping the same region twice, and to retrieve as required what seemed worthy of note.

To make this possible, the movement of the telescope is transmitted by means of a cord ... to the interior of a little observatory built in wood and incorporated in the same moveable framework as the mounting of the telescope. The cord winds round a cylindrical drum as in a small clock; the axis of the drum has a hand that corresponds to a dial divided into 60 parts representing the minutes of a degree in the movement of the telescope up and down, in the sense of its distance from the Pole. A wheel indicates the degrees themselves, so that by glancing at the dial which is on the table where Miss Herschel is writing, one sees continually by the position of the arrow the zone of the heavens corresponding to her brother's telescope; and by the time shown by a pendulum, which part of that zone is currently in the field of view of the telescope. These measurements, which correspond to what astronomers call declination and right ascension, are accurate to one minute of a degree in the former, and two seconds of time in the latter, which is more than enough to avoid any uncertainty. A tube leads down from the gallery where the observer is, and ends in the form of a trumpet located between the two dials. This allows the brother and sister to understand what each other is saying, even when they speak in a low voice.

[*What now follows relates to the 20-ft*] Miss Herschel has in front of her a large book, whose pages are ruled into little squares whose sides measure two lines.

Each zone that has been observed corresponds in this book to a horizontal band four squares in height, and whose width depends on the length of time during which her brother observes. She marks in each square what her brother indicates, and takes a further note. This celestial register is already extensive, and shows at a glance which parts of the heavens have already been swept and which remain to be swept. Mr Herschel hopes to complete the work in ten or twelve years.

The focus of the attention of this incomparable astronomer is drawn less to nearby objects such as the Moon and the planets, than to the depths of Creation, which he seeks to fathom with his powerful tools. So for example he has gained insights into the probable (and at present unknown) distances of the nearer stars. He has established what was previously no more than suspected, that the Milky Way, that whitish and irregular band that circles the sky, is formed of a mass of stars or suns. He has dared to go on to imagine that this complete system, including all the other stars that are visible to us, is nothing other than the system to which our Sun belongs. Then, using his telescope to study the different strata occupied by stars, to a greater or less extent, he has sought to establish the relative position of our Sun in this system and he thinks it placed near one of the extremities. Taking this vast system as itself a unit, he has seen it as nothing more than a tiny part of the Creation known to us, looking on it as a simple nebula; that is, supposing the little whitish nebulosities that are to be seen in various parts of the sky, and of which the most powerful telescopes can make out no details, are made up of assemblages just like our Milky Way plus the stars individually visible to us; and that our system seen from a nebula would appear just like the nebula appears to us....

Published in sections in *Journal de Genève*, 1787, nos. 11–14, 45–62.

In October 1793 the Astronomer Royal, Nevil Maskelyne, visited the Herschels and afterwards wrote an account of how Caroline swept for comets:

I paid Dr & Miss Herschel a visit 7 weeks ago. She shewed me her 5-ft Newtonian telescope made for her by her brother for sweeping the heavens. It has an aperture of 9 inches, but magnifies only from 25 to 30 times, & takes in a field of 1°49′ being designed to shew objects very bright, for the better discovering any new visitor to our system, that is Comets, or any undiscovered nebulae. It is a very powerful instrument, & shews objects very well. It is mounted upon an upright axis, or spindle, and turns round by only pushing or pulling the telescope; it is moved easily in altitude by strings in the manner Newtonian telescopes have been used formerly. The height of the eye-glass is altered but little in sweeping from the horizon to the zenith. This she does and down again in 6 or 8 min, & then moves the telescope a little forward in azimuth, & sweeps another portion of the heavens in like manner. She will thus sweep a quarter of the heavens in one night. The Dr has given her written instructions how to proceed, and she knows all the nebulae [listed by Messier] at sight, which he esteems necessary to distinguish new Comets that may appear from them. Thus you see, wherever she sweeps in fine weather nothing can escape her.

Letter of Nevil Maskelyne to Nathaniel Pigott, 6 December 1793, Royal Astronomical Society Nathaniel Pigott archives.

In November 1794, the Herschels were visited by Giambattista Rodella *(1749–1834), mechanic at the Padua Observatory:*

Another time Secretary Vignola took me to Windsor, thirty or forty miles from London, to see the famous Mr Herschel. He is an excellent person, he treated me with courtesy, gave me a full demonstration of his telescopes and his other artefacts, because he also has a workshop and workmen. I am full of admiration for this remarkable man, not only because of his industry in building innumerable telescopes, but above all for his ingenuity in mounting them and manoeuvring them single-handedly with great rapidity.

From Rodella's *Relazione di Varie Osservazioni in Proposito di Macchine*, reprinted in *La Voce di Hora*, no. 31 (December 2011), 43 *et seq.*, pp. 50–51.

Source Materials

William Herschel's published papers (with one trivial exception) appeared in *Philosophical Transactions of the Royal Society* and are reprinted in *The Scientific Papers of Sir William Herschel*, ed. by J. L. E. Dreyer (2 vols, Royal Society and Royal Astronomical Society, London, 1912). The major cosmological papers are reprinted with commentary and notes in Michael Hoskin, *The Construction of the Heavens: The Cosmology of William Herschel* (Cambridge University Press, Cambridge, 2012).

Caroline's diaries and letters are extensively cited in Mrs John Herschel, *Memoir and Correspondence of Caroline Herschel*, 2nd edn (John Murray, London, 1879); and a granddaughter of William, Constance A. Lubbock, drew on the family archives for her *The Herschel Chronicle: The Life-story of William Herschel and his Sister Caroline Herschel* (Cambridge University Press, Cambridge, 1933, reprinted The Herschel Society, Bath, 2000). Caroline tells her own life-story in *Caroline Herschel's Autobiographies*, ed. by Michael Hoskin (Science History Publications (U.K.), Cambridge, 2003); the first, written when she was in her early 70s, takes her story until the day of William's marriage, while the second, written twenty years later, ends in 1777.

The extensive Herschel archive held by the Royal Astronomical Society and containing most of the scientific papers and correspondence of William and Caroline, along with many papers of John, is available from the Society on both DVDs and CDs.

M. Hoskin, *William and Caroline Herschel*, SpringerBriefs in Astronomy,
DOI: 10.1007/978-94-007-6875-8,

Further Reading

Alexander, A.F.O'D. (1962) *The Planet Saturn: A History of Observation, Theory and Discovery* (Faber & Faber, London), chaps. 8–9

Alexander, A.F.O'D. (1965) *The Planet Uranus: A History of Observation, Theory and Discovery* (Faber & Faber, London), chaps. 1–4

Bennett, J.A. (1976) "'On the Power of Penetrating into Space': The Telescopes of William Herschel", *J. Hist. Astron.*, **7**, 75–108

Bennett, J.A. (1982) "Herschel's Scientific Apprenticeship and the Discovery of Uranus", in G. E. Hunt (ed.), *Uranus and the Outer Planets* (Cambridge University Press, Cambridge), 35–53

Buttmann, Günther (1974) *The Shadow of the Telescope: A Biography of John Herschel* (James Clarke & Co., London)

Chapman, Allan (1993) "An Occupation for an Independent Gentleman: Astronomy in the Life of John Herschel", *Vistas in Astronomy*, **36**, 71–116

Clerke, Agnes M. (1895) *The Herschels and Modern Astronomy* (Cassell & Co., London)

Dreyer, J.L.E. (ed.) (1912) *The Scientific Papers of Sir William Herschel* (2 vols, The Royal Society and The Royal Astronomical Society, London)

Hanham, Andrew, and Michael Hoskin (2013) "The Herschel Knighthoods: Facts and Fiction", *J. Hist. Astron.*, **44**, 149–160

Herschel, Mrs John (1879) *Memoir and Correspondence of Caroline Herschel*, 2nd edn (J. Murray, London)

Hingley, P.D. (1998) "The Will of Sir William Herschel", *Astronomy & Geophysics*, **39(3)**, 7

Hoskin, Michael (1966) "Stellar Distances: Galileo's Method and Its Subsequent History", *Indian J. Hist. Sci.*, **1**, 22–9

Hoskin, Michael (1970) "The Cosmology of Thomas Wright of Durham", *J. Hist. Astron.*, **1**, 44–52

Hoskin, Michael (1977) "Newton, Providence and the Universe of Stars", *J. Hist. Astron.*, **8**, 77–101

Hoskin, Michael (1979) "Goodricke, Pigott and the Quest for Variable Stars", *J. Hist. Astron.*, **10**, 23–41

Hoskin, Michael (1980) "Herschel's Determination of the Solar Apex", *J. Hist. Astron.*, **11**, 153–63

Hoskin, Michael (1992) "John Herschel and Astronomy: A Bicentennial Appraisal", *John Herschel 1792–1992*, ed. by B. Warner (Taylor & Francis, Cape Town), 1–17

Hoskin, Michael (2002) "Caroline Herschel: Assistant Astronomer or Astronomical Assistant", *History of Science*, **40**, 425–44

Hoskin, Michael (2003a) "Herschel's 40ft Reflector: Funding and Functions", *J. Hist. Astron.*, **34**, 1–32

Hoskin, Michael (2003b) "Vocations in Conflict: William Herschel in Bath, 1766–1782", *History of Science*, **41**, 315–33

DOI: 10.1007/978-94-007-6875-8,

Hoskin, Michael (2003c) *The Herschel Partnership: as Viewed by Caroline* (Science History Publications, Cambridge)

Hoskin, Michael (2003d) *Caroline Herschel's Autobiographies* (Science History Publications, Cambridge)

Hoskin, Michael (2004a) "Was William Herschel a Deserter?", *J. Hist. Astron.*, **35**, 356–8

Hoskin, Michael (2004b) "Alexander Herschel: The Forgotten Partner", *J. Hist. Astron.*, **35**, 387–420

Hoskin, Michael (2005a) "Caroline Herschel as Observer", *J. Hist. Astron.*, **36**, 373–406

Hoskin, Michael (2005b) "Unfinished Business: William Herschel's Sweeps for Nebulae", *History of Science*, **43** (2005), 305–20

Hoskin, Michael (2006a) "Caroline Herschel's Catalogue of Nebulae", *J. Hist. Astron.*, **37**, 251–5

Hoskin, Michael (2006b) "Caroline Herschel's Revenge", *J. Hist. Astron.*, **37**, 109–10

Hoskin, Michael (2007) *The Herschels of Hanover* (Science History Publications, Cambridge)

Hoskin, Michael (2008a) "George III's Purchase of Herschel Reflectors", *J. Hist. Astron.*, **39**, 121–4

Hoskin, Michael (2008b) "Nebulae, Star Clusters and the Milky Way: From Galileo to Herschel", *J. Hist. Astron.*, **39**, 363–96

Hoskin, Michael (2010a) "Mary Herschel's Fortune: Origins and Impact", *J. Hist. Astron.*, **41**, 213–23

Hoskin, Michael (2010b) "William Herschel and the Southern Skies", *J. Hist. Astron.*, **41**, 503

Hoskin, Michael (2011a) "William Herschel and Herschelian Reflectors", *History of Science*, **49**, 115–20

Hoskin, Michael (2011b) "William Herschel and the Nebulae", *J. Hist. Astron.*, **42**, 177–92, 321–38

Hoskin, Michael (2011c) *Discoverers of the Universe: William and Caroline Herschel* (Princeton University Press, Princeton)

Hoskin, Michael (2012a) "William Herschel's Residence in Bath, 1799–1801", *J. Hist. Astron.*, **43**, 351–8

Hoskin, Michael (2012b) "William Herschel's Agenda for His Son John", *J. Hist. Astron.*, **43**, 439–54

Hoskin, Michael (2012c) *The Construction of the Heavens: The Cosmology of William Herschel* (Cambridge University Press, Cambridge)

Hoskin, Michael (2013) *Caroline Herschel: Priestess of the New Heavens* (Science History Publications/USA, Sagamore Beach, Mass.)

Hoskin, Michael, and David W. Dewhirst (2006) "William Herschel and the Prehistory of Stellar Spectroscopy", *J. Hist. Astron.*, **37**, 393–403

Hysom, E.J. (1996) "Tests of the Shape of Mirrors by Herschel", *J. Hist. Astron.*, **27**, 349–52

Lessens, Ronald (2004) *William Herschel: Musicien Astronome* (Burillier, Vannes)

Lubbock, Constance A. (1933) *The Herschel Chronicle: The Life-story of William Herschel and his Sister Caroline Herschel* (Cambridge University Press, Cambridge)

Maurer, Andreas (1998) "A Compendium of All Known William Herschel Telescopes", *Journal of the Antique Telescope Society*, no. 14, 4–15

Porter, Roy (1982) "William Herschel, Bath, and the Philosophical Society", in G. E. Hunt (ed.), *Uranus and the Outer Planets* (Cambridge University Press, Cambridge), 23–34

Schaffer, Simon (1980) "'The Great Laboratories of the Universe': William Herschel on Matter Theory and Planetary Life", *J. Hist. Astron.*, **11**, 81–111

Schaffer, Simon (1981) "Uranus and the Establishment of Herschel's Astronomy", *J. Hist. Astron.*, **12**, 11–26

Spaight, John Tracy (2004) "'For the Good of Astronomy': The Manufacture, Sale, and Distant Use of William Herschel's Telescopes", *J. Hist. Astron.*, **35**, 45–69

Steavenson, W.H. (1924–25) "A Peep into Herschel's Workshop", *Transactions of the Optical Society*, **26**, 210–37

Turner, A.J. (1977) *Science and Music in 18th Century Bath* (University of Bath, Bath)
Warner, Brian (1992) "Sir John Herschel at the Cape of Good Hope", in *John Herschel 1792–1992*, ed. by B. Warner (Taylor & Francis, Cape Town), 19–55

Index

DOI: 10.1007/978-94-007-6875-8,

MIX
Papier aus verantwortungsvollen Quellen
Paper from responsible sources
FSC® C105338

If you have any concerns about our products,
you can contact us on
ProductSafety@springernature.com

In case Publisher is established outside the EU,
the EU authorized representative is:
Springer Nature Customer Service Center GmbH
Europaplatz 3, 69115 Heidelberg, Germany

Printed by Libri Plureos GmbH
in Hamburg, Germany